Handbook on
Striped Murrel, *Channa striata* (Bloch, 1793)
Reproductive Biology, Breeding, Seed and Grow-out Productions

About the Authors

Dr. Sudhir Raizada (M.Sc., Ph.D., LL. B) did his master's and doctorate degrees from the Agra University, Agra in Zoology (Fish and Fisheries). He joined ICAR-Central Institute of Fisheries Education (Deemed University), Mumbai in 1979 as Scientist and reached the position of Principal Scientist in the discipline of fish and fisheries science. He joined ICAR-National Bureau of Fish Genetic Resources, Lucknow in 2009 and worked for the conservation of fisheries. He was selected to the position of Assistant Director General (Inland Fisheries) at Indian Council of Agricultural Research (Ministry of Agriculture and Farmers' Welfare, Govt. of India), New Delhi in 2016 and superannuated from there in 2018.

Presently, he is engaged with Emeritus Scientist Scheme of ICAR on the topic "Up scaling protocols for brood stock development, breeding and seed production of striped murrel (*Channa striatus*)". Dr Raizada has been involved in teaching/research/extension in the field of Aquaculture and Fisheries Resources Management for more than 42 years. He has successfully bred more than 25 fish species such as major carps, minor carps, catfishes, murrels and freshwater prawn. First time, the technology of seed production of giant freshwater prawn (*Macrobrachium rosenbergii*) and commercial farming of tiger shrimp (*Penaeus monodon*) using inland saline water was developed and demonstrated by his team under his leadership in Haryana. He has also worked on modern aquaculture farming systems such as RAS and BFT. He has handled more than 25 research projects and published 63 peer-reviewed scientific papers, 7 books, 6 book chapters, 25 manuals and 50 popular articles. Ten students completed their Master's Dissertation and three Ph.D. Thesis under him. He has been awarded ICARs Dr. Rajendra Prasad Biannual Award for his Hindi book "Bhartiya Maatyaki" and 3 institutional awards as a Best Researcher at CIFE and NBFGR. (CV enclosed as Annexture-I)

Dr Prem Prakash Srivastava, Principal Scientist did his Ph.D. degree in Biochemistry from the University of Allahabad and joined the ICAR-Central Institute of Fisheries Education, Mumbai (Deemed University) in 1990 in the Fish Nutrition and Biochemistry Laboratory. He is a well recognized scientist in fish nutrition with more than 150 scientific publications. He has tirelessly supported establishment of fish nutrition and nutritional biochemistry academic and other institutions in the country also. He has coordinated large number of multi-disciplinary research projects on fish nutritional biochemistry of carp, catfishes, and murrels. He has been honoured with several awards viz. Bursary Scholarship in M.Sc.(Biochemistry) for 2 year (81-83), University Silver Medal; Justice Harish Chandra Gold Medal; Young Scientist Award-Swarn Jayanti Puruskar; Dr. Heera Lal Choudhary Team Award-two times; Dr. Zahoor Qasim Gold Medal;

Best Technical Award; Young Scientist Associate Award; Fellow of the Academy of Science, Engineering and Technology (FASET); Best Scientist Award; Appreciation Award; Bioved Fellowship Award; Best Extension Scientist; Prof. D.K. Belsare Medal etc. He got first class Diploma in Computer Programming (DCP) from S.P. Jain Institute of Management and Research, Bhavan's College, Mumbai and P.G. Diploma in Bioinformatics from BII, Noida. He has teaching experience of 32 years and research experience of 34 Years. He has availed post – doctoral DBT Overseas fellowship at Nutrition, Aquaculture & Genomics, INRA, Saint-pee-sur-Nivelle, France. Presently, working as Registrar, Dr Rajendra Prasad Central Agricultural University (Govt. of India), Samastipur, Bihar – 848 125, India.

Dr. Anurag Rawat did his doctorate degree from the School of Life Sciences, Babasaheb Bhimrao Ambedkar Central University, Lucknow. He has been awarded the positions of JRF, SRF and currently engaged as Research Associate (RA) at ICAR- National Bureau of Fish Genetic Resources, Lucknow. Dr Rawat has 10 years of experience in fish reproductive biology, broodstock rearing, fish breeding and larval rearing of catfishes and murrels. He has published 13 peer reviewed scientific papers in various journals, a book chapter and a paper in souvenir. He has presented 34 papers in various seminar, workshop and conference and has been conferred with best poster paper award twice. He has also served as Lecturer Zoology in Santosh Kumar P.G. College, Hardoi, U.P. for 4 years.

Handbook on
Striped Murrel, *Channa striata* (Bloch, 1793)
Reproductive Biology, Breeding, Seed and Grow-out Productions

Sudhir Raizada
Emeritus Scientist
and
Former Assistant Director General (ICAR)
Indian Council of Agricultural Research
New Delhi

Prem Prakash Srivastava
Principal Scientist and
Presently Registrar
Central University Rajendra Prasad Agricultural University
Samastipur-848 125, Bihar

Anurag Rawat
Research Associate
ICAR-National Bureau of Fish Genetic Resources
Lucknow, Uttar Pradesh

NEW INDIA PUBLISHING AGENCY
New Delhi-110 034

NEW INDIA PUBLISHING AGENCY
101, Vikas Surya Plaza, CU Block, LSC Market
Pitam Pura, New Delhi – 110 034, India
Email: info@nipabooks.com
Web: www.nipabooks.com

For customer assistance, please contact
Phone: + 91-11-27 34 17 17 Fax: + 91-11- 27 34 16 16
E-Mail: feedbacks@nipabooks.com

ISBN: 978-93-90512-87-4

Composed and Designed by NIPA

Preface

The rapid growth of aquaculture in recent has raised questions concerning its environmental sustainability. To minimize the potential threat for climate change, there is need to modernize the sector in a manner that is more responsible, sustainable, & transparent and should generate enhanced public and private confidence in aquaculture. Therefore future aquaculture should be targeted to produce more fish using less water or recycle the water through elimination of harmful effluent discharge using intensive farming practices in open as well in closed systems. Whereby closed-systems can rely on modern systems of re-circulatory aquaculture (RAS), biofloc Technology (BFT) and aquaponics; open systems of farming in ponds, cages and pens will require mechanical aeration and water quality management protocols for sustainable growth potential. Present day freshwater aquaculture is mainly carp centric, a low value fish and hence operating modern systems of farming with carps may not be able to substantiate outputs. Further, carps are not suitable to grow in small enclosures due to growth problems. In order to effectively use these systems, the focus now should be on aquaculture of high value species, which can sustain intensive stocking with minimum disease risk.

Striped murrel or striped murrel or chevron murrel (Scientific name, *Channa striata*) is an important species of many countries of South-east Asia including India. It is popular for its esteemed taste, therapeutic values, antinociceptive properties and hence consumed by many to induce healing and pain control after clinical operations. It contains high arachidonic acid (AA), which is a precursor of prostaglandin that may initiate blood clotting and enhance tissue growth. It is devoid of muscular spines and hence easily processed for making fillets for meeting the export demand. The species is air-breathing and can live without water for few hours and with little water for several hours, hence suitable for marketing in live condition to nearby and far places. In comparison to carps, the growth of this species is better in small rearing enclosures and hence farmers having small landholdings and entrepreneurs who wish to adopt intensive farming under indoor conditions, can make great use of this high value farming species.

The efforts made in the past on farming of this species has not shown promising results due to several problems associated with its breeding, seed production, larval rearing, grow-out production and non-availability of species and size-specific artificial feeds. However major handicaps are now being understood with recent global research and it has become possible to overcome them to a larger extent. Therefore efforts have been made to compile the outcome of available research with personal research experience of the authors to provide updated information on this important valuable species for the promotion of aquaculture in respect of reproductive biology, breeding, larval rearing, cannibalism mitigation issues, fingerling rearing and grow-out productions under different farming system.

The authors wish that the book will be highly useful to the researchers, students, entrepreneurs and farmers in better understanding various issues relating to seed production and farming of striped murrel and help the farmers in adopting this high value species for its farming. It is hoped that the propagation of this species will increase our efforts towards species diversification and meeting the call of our "Honble Prime Minister for doubling farmers" income. Protocols suggested will also help the stakeholders in propagating other species of this genus, particularly those having ornamental importance.

Sudhir Raizada
Prem Prakash Srivastava
Anurag Rawat

Acknowledgements

This book contains the outcome of research work carried out by the authors and other researchers for the promotion of farming of a highly valuable fish species, striped murrel (*Channa striata*). However, this has become possible with the kind support of many persons, who helped us in carrying out work on this important fish species and encouraged us to write this book and hence authors wish to share their gratitude to them. At the foremost, the authors would like to express their special thanks of gratitude to Dr Trilochan Mohapatra, Secretary DARE, Government of India and the Director General, Indian Council of Agricultural, Research New Delhi for according approval to the Emeritus Scientist Scheme entitled "Up-scaling protocols for brood stock development, breeding, larval rearing, and fingerling production of striped murrel Channa striatus, for species diversification and conservation" under which upscaling of the protocols defined in this book have been developed. We are also extremely grateful to Dr R.C. Agarwal, Deputy Director General (Education Div.), ICAR, New Delhi for providing all support in successfully operating the scheme. The authors also over helmed in all humbleness and gratefulness to acknowledge Dr. J.K. Jena, Deputy Director General (F.S.), ICAR for encouragement and providing useful suggestions for incorporation in the research. We express our heartfelt gratitude to Dr Kuldeep Lal, Director, National Bureau of Fish Genertic Resources, Lucknow for providing both field and wetlab facilities for carrying out the work. First two authors are extremely grateful to the financial support provided by the Indian Council of Agricultural Research, New Delhi under the Emeritus Scientist Scheme, which helped in up-scaling the protocols and writing of this book. We also thank Dr M.K. Agnihotri, Assistant Director General (HRD), Education Division, ICAR for his timely support in the operation of the scheme. The authors also complement and thank all the researchers, whose work have been incorporated in this book to provide global information for wider knowledge. We also thank all those workers who supported this work in any way, naming few of them Dr Vikash Sahu, Dr. Kailash Chandra Yadav, Dr. Ajay Singh and Dr Rajesh Dayal for their excellent support. We will fail in our duty if we don"t acknowledge the support of our family members, which made possible to conduct work on this species and helped us in bringing out this book.

Sudhir Raizada
Prem Prakash Srivastava
Anurag Rawat

Acknowledgments

Contents

1

Introduction

The fish, Striped murrel; taxonomic name, *Channa striata* is known by various names, prominent in scientific domain beings, striped murrel, Chevron murrel in English and Saul, shawl, shol, Haruan (Malaysia) in local languages. It is classified as under:

Phylum	: Chordata
Sub-Phylum	: Vertebrata
Super class	: Gnathostomata
Class	: Actinopterygii
Order	: Perciformes
Family	: Channidae
Genus	: *Channa*
Species	: *striata*.

The species is commonly found in tanks, ponds, lakes, reservoirs, canals, swamps, etc in freshwater plains. It is air-breathing in nature and survives easily low oxygen stress. During drying-up of water bodies in summers, it has been reported to penetrate deep under moist-earth and come out live on arrival of rains. The name 'murrel' encompasses its resemblance to reptile (snake) due to presence of body scales all over its body including head region. Fully-grown adults are dark greenish brown with faint black and white vertical bands visible across its entire abdominal length below the lateral line and a black ocellus at posterior end of dorsal fin.

Striped murrel is one of the important food fish of high commercial value due to preference for its meat quality, therapeutic and antinociceptive properties and hence consumed by many to induce healing and pain control after clinical operations. It contains high arachidonic acid (AA), which is a precursor of prostaglandin that may initiate blood clotting and enhance tissue growth. Geographically, the species is distributed in Western to Southeast Asia including Bangladesh, Cambodia, China, India, Indonesia, Lao People's Democratic

Republic, Malaysia, Myanmar, Nepal, Pakistan, SriLanka, Thailand and Vietnam (Chaudhary, 2019). It is supposed to be introduced in countries like Fiji, New Caledonia (Welcomme, 1988), Madagascar (Stiassny and Raminosoa, 1994) and Hawaii (Courtenay *et al.,* 2004). The species has been reported to grow to a length of 100 cm (Rahman, 1989) and categorized,, *Least Concern* under the IUCN Red List (*IUCN, 2020*).

Globally, 92523 ton of striped murrel is produced (FAO, 2016), of which the major share of 70802 ton is contributed from capture fisheries and 21721 ton from aquaculture (Figure 1). Though capture fisheries of striped murrel is quite stable in terms of global production, the aquaculture has shown a steady rise in last two decades. In India, though no production data is available, the production of this species is mainly contributed from the capture fisheries with meager contribution from aquaculture. This has caused serious stress in past few years on its resources, more particularly due to habitat degradation and man-made hazards in aquatic ecosystem.

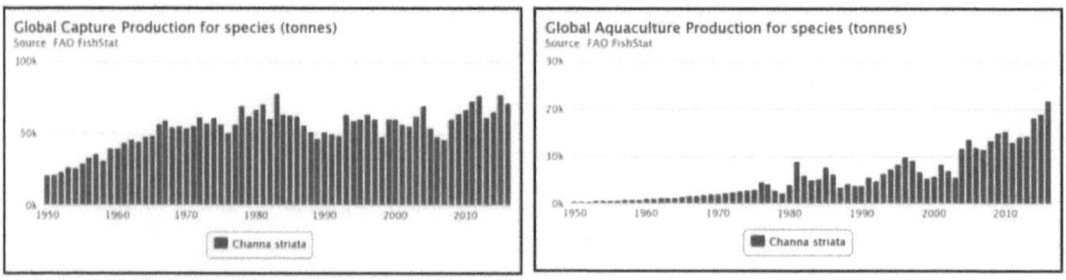

Fig. 1: Global capture and aquaculture production of striped murrel (FAO 2020)

With burgeoning market demand and declining natural stocks, the species needs serious efforts to sustain its fisheries both through natural stock enhancement programs and aquaculture to meet the internal market demand as well export of this species, which is highly suitable for making high quality fish-fillet greatly preferred abroad (Ayyappan *et al.*, 2001). The efforts made in the past on farming of this species has not shown promising results due to several problems associated with its breeding, seed production, larval rearing, grow-out production and non-availability of species and size-specific artificial feeds. However with recent global research, it has become possible to overcome the above issues to a larger extent. Therefore efforts have been made to compile the outcome of various research works carried out by different workers including that of authors of this book in respect of reproductive biology, captive breeding, seed and grow-out production of this valuable species in order to provide updated information for the promotion of aquaculture.

2

Reproductive Biology

Knowledge on fish reproductive biology is extremely important to understand various issues relating to reproduction and fishery management. Reproductive success in aquaculture systems depends to a greater extent on our knowledge about sexual differentiation, gonad structure, gametogenesis, spawning, larval metamorphosis, etc for successful captive breeding of a fish species. This chapter therefore, contains information on secondary sexual characters, sex-ratio, gonad structure, gonadal maturity, fecundity and embryonic development of striped murrel.

2.1 Secondary Sexual Characters

The sexes are separate in striped murrel however differentiation of sex is not easy when fish is immature. The differentiations in sexes become distinct with advancement of maturity during spawning season. On attaining full maturity, a good fecund mature female could be easily distinguished with having bulged belly originating nearing the site of pectoral fins and extends near to the vent.

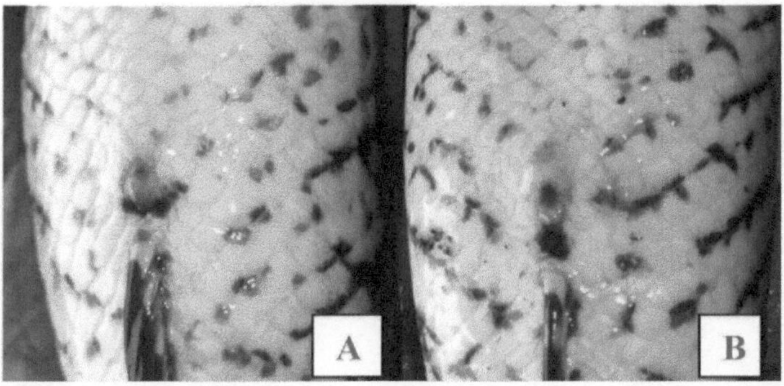

Fig. 2: (A) Gonopore of female with swollen vent and no space between vent and anal fin. **(B)** Gonopore of male with anal papilla and space between pailla nd anal fin

The gonadal orifice of females is pear-shaped, fleshier, pinkish and almost touching the origin of anal fin. Females do not have genital papilla even when fully mature (Fig. 2A). Males on the contrary have slender body, no bulge abdomen, elongated genital orifice with little fleshy mass and have minute genital papilla at the time of complete maturity. The space between genital orifice and anal fin origin is distinct (Fig. 2B). Generally, males are darker in body colour when fully mature (*Devraj*, 1973; *Parmeswaran & Murugesan*, 1976a; *Rawat et al.,* 2020). The males do not exude milt on applying pressure over the abdomen as in case of carps and many other fishes.

2.2 Gonad Structure

Testes: Males have roughly tubular, bilobed testes with left lobe slightly larger. The size of testes is very small in comparison to carps and catfishes and sometime difficult to locate in immature fishes. However, it extends in size with advancement of maturity. Two lobes of the testes joined before entering the cloaca, where a third lobe is also formed. The anal part of testes in striped murrel remains attached with the ventral wall mesenteries of the coelomic cavity and then finally gets lying along with the intestine. Testes is off-white to brownish red and reddish white during immature and maturing phases, but change to yellowish-orange in the mature phase (So *et al*., 2012 and Rawat *et al*., 2020) (Fig. 3A, B, C, D). Newhard (2015) has described the location of testes in C. argus body similar to the above work.

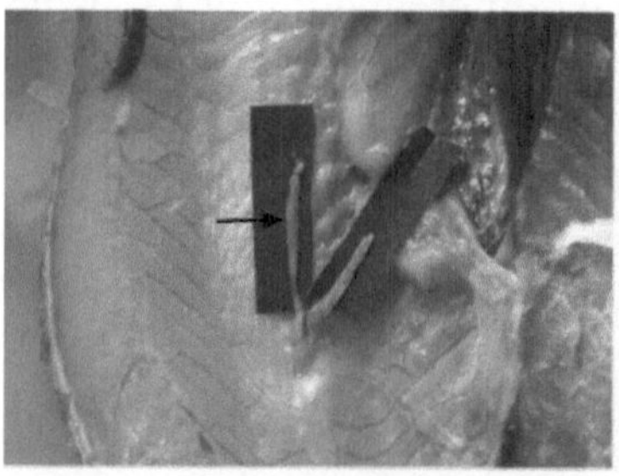

A. Immature Phase of Testes *in situ*

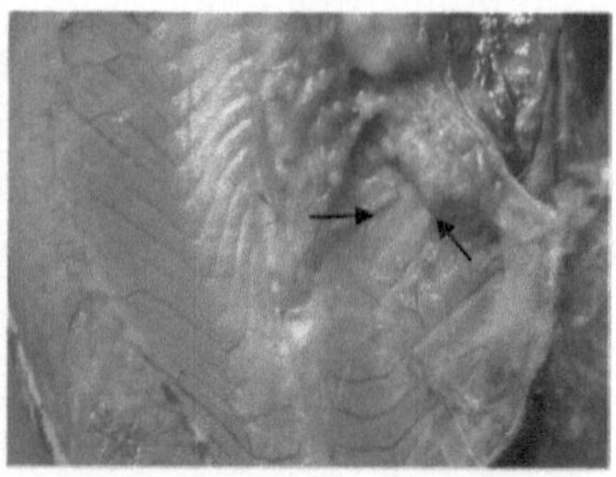

B. Maturing Phase of Testis in situ

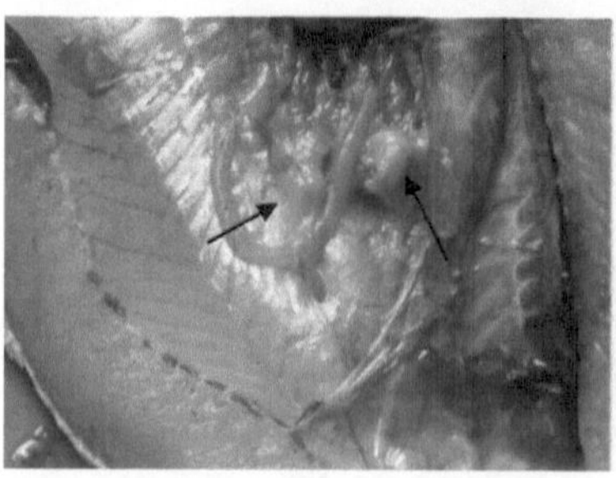

C. Mature Testes *in situ*

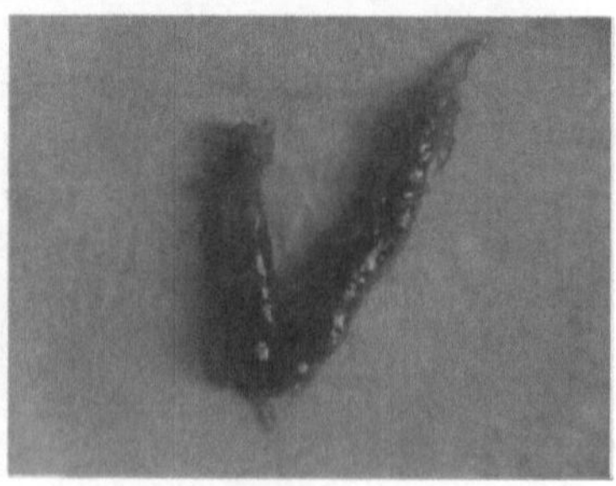

D. Extracted Testes

Fig 3: Developing stages of testes in striped murrel (*Channa striata*)

Ovary: The ovary of striped murrel is also a paired organ and situated in the coelomic cavity and suspended on either side of the mid-line by a mesovarium. The ovary is bilobed, each lobe is spindle-shaped and varies in size with left lobe being larger in length than the right one. Initially it is small, and looks like a muscular reddish mass in colour and without oocytes (Fig. 4A). With advancement of maturity, the size of the ovary increases both in length and diameter and oocytes become visible in the ovarian follicle. Initially, oocytes are pale yellow in colour (Fig. 4B). With further ovarian development, the length and size of both lobes of ovary increases and the ovarian follicles are found to fill entirely with the oocytes, which are pale yellow in colour. The ovarian follicles at this stage had large network of blood capillaries over its outer surface. With further advancing of maturity, the ovarian lobes increase more in diameter than the length due to increase in the number and size of oocytes and both lobes of ovarian follicles become flaccid. The colour of ova now turns to orange and there was reduction of network of blood capillaries over the ovarian follicles, which marked attaining of complete maturity in the fish (Figure 4C & 4D). Besides distinct ovarian and testicular structures, ovotestis (intersex) have also been reported in striped murrel in length range 81.00-229.00mm, symbolizes that the species exhibit hermaphroditism in some of the specimens (Irmawati *et al.*, 2019).

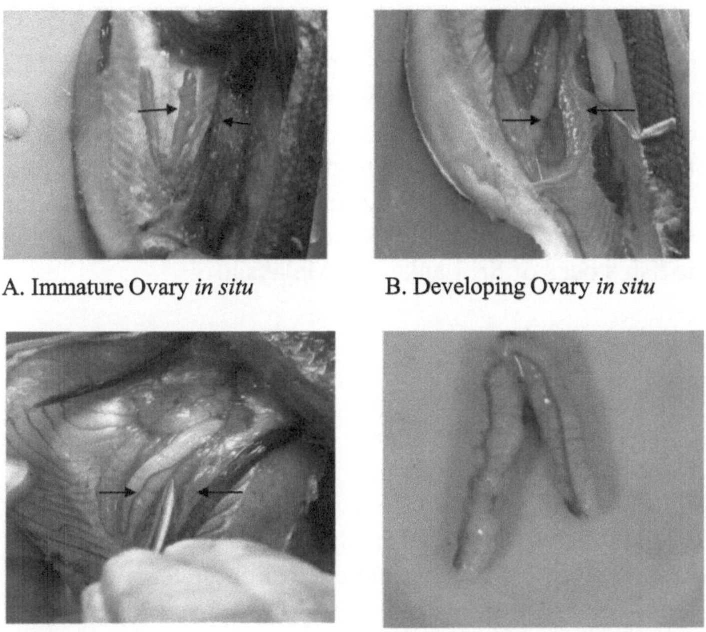

A. Immature Ovary *in situ*

B. Developing Ovary *in situ*

C. Developed Ovary *in situ*

D. Developed Ovary after extraction

Fig. 4: Developing stages of Ovary in striped murrel (*Channa striata*)

2.3 Sex Ratio, Gonad Maturity and Fecundity

2.3.1 Sex Ratio

Remarkably, lots of variations in the sex ratio in murrels have been reported. So *et al.* (2012) have observed significantly low male to female sex-ratio in wild collections ranging from 1:2 to 1:3 in striped murrel. However, there is contradictory report from Morioka *et al.* (2016), who did not observe much variation in sex ratio during most part of the year. The variations in sex-ratio in other murrel species *Parachanna obscura* and *C. punctata* have been reported as 1:4 (Olurin and Salvage, 2011) and 1:2 (Koundal *et al.*, 2014) respectively. Besides low ratio of female to male population in natural waters, the other features of concern are early maturity in males and maturity attainment by only few male brooders in a mixed sex population at a time in a tank/pond. The authors are of the view that low male sex ratio and suppression of male maturity in murrels in a water body may occur due to some sort of anti-cohort linkage dominated by male aggression in their jurisdiction area over the recessive males. However, this needs to be verified in future research.

2.3.2 Size/Age at Maturity

Striped murrel attains maturity in the first year of its life. The average mixed-sex first maturity in striped murrel has been reported at a length of 11.56 cm to 23.00 cm (Moriaka *et al.*, 2016; Irmawati *et al.*, 2019) in river and wetlands. The ovarian development starts at 11.56±1.96 cm, while testicular at 21.00±4.24 cm (Irmawati *et al.*, 2019). Female striped murrel has been reported to attain complete maturity at 26.45±3.07 cm in total length and 167.4±48.09 g in body weight. Similar recordings have also been made by Kilambi (1986) and Ali (1999) in the fish stocks of Sri Lanka and Malaysia.

2.3.3 Gonado-somatic Index

The gonads are comparatively small in striped murrel. The female striped murrel has been reported to have gonadosomatic index (GSI) of 3.69 to 11.5 percent during May-August months, with maximum being at the time of spawning (Alikunhi, 1953; Murugesan and Kumaraiah, 1976; Pati *et al.*, 2004; Thumronk *et al.*, 2011; Ferdausi *et al.*, 2015; Mahmod *et al.*, 2016;; Yadav, *et al.*, 2016). Morioka *et al.*, (2016) have obtained highest female GSI of about 4 percent in April, the main spawning season of this species in Laos, while Ghaedi *et al.* (2013) recorded 11.5 percent during August in Malaysia. Yadav *et al.* (2016) estimated GSI of female fishes from immature to mature phases during January to June months and found them to range 0.274-4.32 percent. In case of male fishes, the GSI is very low and observed between 0.030-0.106 percent all through the year. Yadav *et al.* (2016) recorded male GSI from 0.138 to 0.230 percent during January to June months being the immature to mature phases of maturity. Mahmod *et al.* (2016) recorded GSI of male fishes to be 0.14 percent during July, being the peak spawning season.

2.3.4 Ovarian Development

The ovary passes through four (Mahmud *et al.*, 2016) or five (So *et al.*, 2012) ovarian developmental stages: immature, maturing, mature, spawning, and post spawning. Microscopic examination and the oocytes size distribution of ripe ovaries indicated that striped murrel exhibits an asynchronous ovarian growth pattern, with all type of oocytes present during the spawning season, reflecting a long spawning period of this species (Table 1 & Figure 5). In asynchronous oocyte development, egg release is concomitant with oocyte recruitment, and therefore ovarian weight only changes slightly during the spawning season, which is similar to Sparidae species (*Dentex dentex*) and red-porgy (*Pagrus pagrus*) respectively (Kokokiris *et al.*, 2001; Ismail *et al.*, 2016).

Table 1: Microscopic description of the maturity stages of the ovaries of striped murrel

Stage of ovary	Characteristics of oocytes
Stage I (Immature)	Oogonia, chromatin nucleolar and early perinucleolar oocytes are present. The perinucleolar oocytes have a large nucleus.
Stage II (Maturing/ Pre-spawning)	Partially yolked oocytes and late perinucleolar oocytes were present. Partially yolked oocytes contain yolk vesicles in their cytoplasm.
Stage III (Mature)	Advanced yolked oocytes are abundantly present. The cytoplasm is filled with yolk granules.
Stage IV (Spawning)	Ovary filled mostly with advanced yolked oocytes with moving of germinal vesicle to one end.
Stage V (Spent/Recovering)	Appearance of empty follicles and few numbers of oogonia and perinucleolar oocytes represented the reserved stock for further spawning. The spend gonads appeared in August and September.

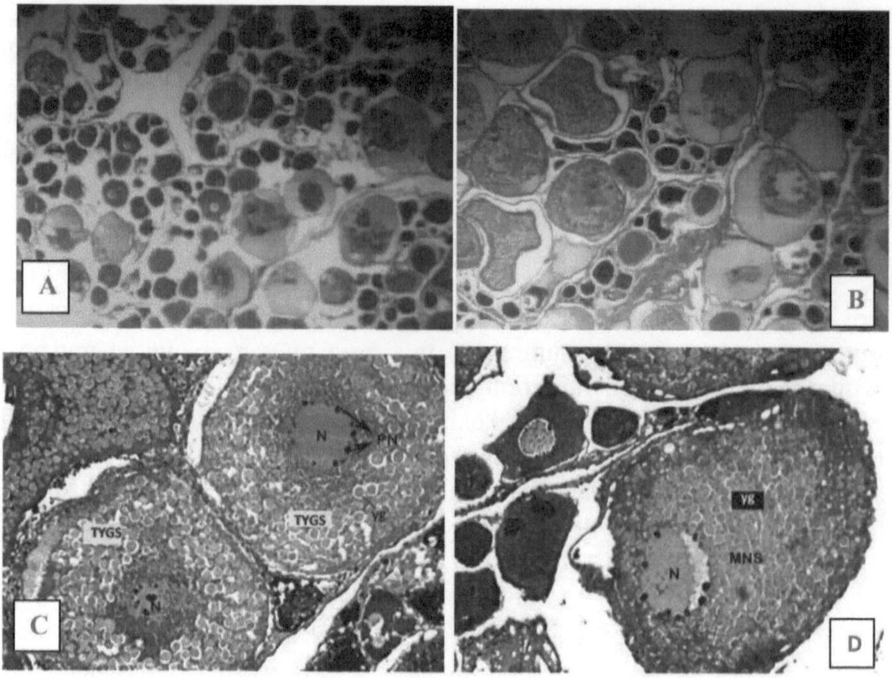

Fig. 5: T.S. of ovary of striped murrel (*C. striata*) **A:** Showing four stages of oocyte development, chromatin nucleolus stage (CNS), early perinucleolus stage (EPS), late perinucleolus stage (LPS), yolk vesicle stage (YVS), indicating immature stage of ovary (Magnification 40x); **B:** Showing four stages of oocyte development, chromatin nucleolus stage (CNS), early perinucleolus stage (EPS), late perinucleolus stage (LPS), yolk vesicle stage (YVS), indicating early maturity stage of ovary (Magnification 40X); **C:** Showing oocyte in the tertiary yolk globular stage showing mature oocyte stage of ovary (Magnification 100X); **D:** Showing migrating nucleus stage (MNS) along with yolk globule oocyte indicating mature stage of ovary (Magnification 100X).

The process of oocyte development in striped murrel has been described by Mahmud *et al.* (2016). According to them, oogonia proliferate through mitotic division of primary germ cells and transform into previtellogenic oocytes, characteristic of the immature ovary. The elaboration of yolk in the oocyte marks the beginning of vitellogenesis at the end of which the cell attains its maximum size and undergoes ovulation, followed by the extrusion of egg to the exterior. The mean oocyte diameter of striped murrel during various developmental stages ranged from 280.00±12.80 μm to 1257.50±24.17 μm. Seven oocyte developmental stages have been identified in striped murrel in histological study, which are described as under.

Chromatin nucleolar stage oocytes are small cells with large nucleus, thin indistinct peripheral zone and with one or more nucleolus.

Early perinucleolar stage oocytes are mostly polygonal in shape and having larger cytoplasmic and nuclear volume. Nucleoli are smaller, increased in number and appeared around the periphery of the nucleus.

Late perinucleolar stage oocytes further enlarge in size and attained polygonal to oval shape with enlargement of nucleus.

Yolk vesicle stage is characterized by the formation of small yolk vesicles in the cytoplasm which appeared as empty unstained vacuoles. The yolk vesicles first appeared at the periphery of the oocyte and gradually spread towards the central nucleus.

Early yolk granule stage is characterized by the formation of numerous small yolk protein granules in the cytoplasm. They first appeared in the outer cortex and gradually increased in size and number as they moved towards the inner cortex. Yolk vesicles were eventually no longer apparent in the cytoplasm.

Late yolk granule stage is reached with further increase in diameter of the oocyte. Fusion of yolk granules and lipid droplets occurred to varying extents.

Spent stage ovaries are characterized by the appearance of empty follicle and few numbers of oogonia and perinucleolar oocytes representing the reserved stock for further spawning.

2.3.5 Testicular Development

The development of testes in striped murrel is similar to a teleostean fish. It follows proliferation of spermatogonia through mitotic division of primary germ cells that transformed into spermatocytes, a characteristic of testes. Four maturity stages in the testis of striped murrel have been described (Bhatti, 2010; Mahmud, 2016; Yadav, 2016) as given in Table 2, Immature: seminal lobes totally occupied by clusters of spermatogonia; Maturing or Pre-spawning: spermatocytes constituting >50% of the germ cells & fewer spermatogonia; Mature or Spawning: higher proportion of spermatids and spermatozoa dominating with decreased percentage of the spematocytes and Spent or Post-spawning: No spermatozoa, with sporadic appearance of spermatogonia and spermatocytes.

Table 2: Microscopic description of maturity stages of testes of striped murrel

Stage of testes	Characteristics of testes
Stage I (Immature)	Testis containing many spermatogonia (SG). A fewspermatocytes (SC) are also thinly scattered.
Stage II (Maturing/ Pre-spawning)	Testes filled with spermatocytes and good amount of spermatids (ST).
Stage III (Mature/Spawning)	Testes contains large amount of spermatozoa (SZ).
Stage IV (Spent/ Post spawning)	Testes containing residual spermatozoa.

Histological study of the testes revealed four developmental stages of male gamets viz. spermatogonia, spermatocytes, spermatids and spermatozoa (Mahmud *et al*., 2016) (Fig. 6), which are described as under.

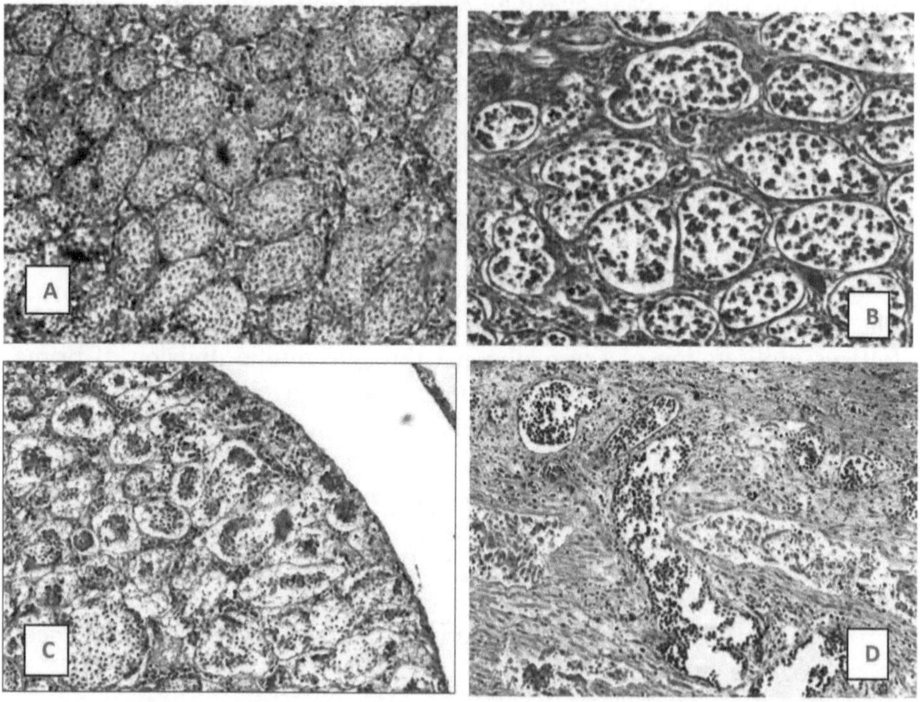

Fig. 6: T.S. of testes of striped murrel (*C. striata*) from cranial part **A:** Showing well organized seminiferous tubules containing spermatocyte in primary and secondary developmental stage (Magnification 40X); **B:** showing advanced stage of spermatogenesis inside the seminiferous tubules, where spermatocytes have been disintegrated from the leydig cells (Magnification 40X); **C:** Cross section depicts Tunica albuginea (outer membrane) of single lobe of testis and spermatozoa in different developmental stages with advance stage moving towards centre of lobule (Magnification 40X); **D:** Showing some seminiferous tubule with spermatozoa, and some empty after release. The vas efference is full of spermatozoa. (Magnification 100X).

Spermatogonia stage comprises of spherical spermatogonia cells with a well defined nucleus and cell membrane.

Spermatocyte stage result due to transformation of spermatogonia into spermatocytes by meiotic division. Spermatocytes are spherical in shape and contain a nucleus in the centre which is smaller in size than the spermatogonia.

Spermatid stage is reached due to transformation of spermatocytes into the spermatids. Spermatid cell are spherical in shape in which nucleus is not clearly seen due to its dark appearance.

Spermatozoa stage are the smallest cells of the spermatogenic lineage appeared as small black colored spots under the microscope. They are the functional male gametes originated from spermatids by the process of spermiogenesis.

2.3.6 Fecundity

The relative and absolute ovarian fecundity of this species has been reported by various workers. Weight-wise relative fecundity of this species has been reported to vary from 29000-55000 kg^{-1} body weight (bw) by *Murugesan and* Kumarariah (1976), 22000-34000 per kg^{-1} bw by Alikunhi (1953) and 95301-112234 per kg^{-1} bw by Pati *et al.* (2004). Ghaedi *et al.* (2013) and Rahman and Awal (2016) have estimated the absolute fecundity of the species in different size groups and recorded it to be 33939±3388 in the size range of 28.02±1.71 cm and 16,330 to 56,467 in the size range of 34.2-51.50 cm. The oocyte diameter varied from 0.20–0.71 mm in stage II to 0.96–1.62 mm in stage IV (Ghaedi *et al.*, 2013). Striped murrel is a batch spawner species and hence ova of different maturity stages in the ovary can be seen at times. However, it releases only mature ova during spawning. Considering the ovarian fecundity reported above by different workers, the numbers of egg spawned by females at a time are comparatively very low. A total of 4050-9500 eggs in the weight range of 260-600g have been reported by Hossain *et al.* (2008).

2.3.7 Embryonic Development

The embryonic development of murrels such as striped murrel (*C. striatus*), giant murrel (*C. marulius*), spotted murrel (*C. punctatus*) have been described by Marimuthu & Haniffa (2007), Thumronk *et al.* (2011), Yadav (2016) and Haniffa *et al.* (2003) in the past. However, Rawat *et al.* (2020) have recently described the embryonic development of striped murrel as under (Table 3, Fig. 7).

In striped murrel, the fertilized eggs are round, translucent, pale yellow, floating and non-adhesive. It has been reported to have diameter of 1.25mm by Willey (1909), 1.04-1.57mm by Ghaedi *et al*., (2013) and 1.50±0.05mm by Rawat *et al*., (2020) (Fig. 7A). Eggs of striped murrel are unique in having single large round oil globule of dia 1.1±0.1 mm, which occupies the greater part of the ovum and immersed in the golden yellow yolk having a dia of 1.25±0.1 mm and both were surrounded within a space bounded by the vitelline membrane. The blastoderm was found situated on the down side of the yolk as a thin layer under the microscope and usually not visible up to morula stage due to high buoyancy of oil globule which always keeps egg upside down (Rawat *et al.,* 2020).

The first cleavage was followed with equal division of blastoderm (2-cell stage) within 20±04 post fertilization minute (pfm). It was followed with second and third cleavages with the formation of 4-cell and 8-cell stages (Fig. 7A) respectively, within 30±05 pfm. The fourth cleavage with formation of 16 cells was observed at 45±05 pfm (Fig. 7B). Further cleavages formed 64 cells in 60±10 pfm. The blastoderm obtained morula stage in 1.45±0.15 pfh (Fig. 7C). The blastula stage was attained in 03.50±0.15 pfh (Fig. 7D). The embryo developed to gastrula (Fig. 7E) and early neurula stages (Fig. 7F) in 05.45±0.15 pfh and 06.30±0.30 pfh respectively. At this stage, head and tail ends become distinct and formation of germ ring is seen. Cephalic region becomes wider with formation of primitive brain, extension of tail region and rudiments of heart could be seen at 10 myotome stage (Fig. 7G). The size of the embryo expanded further with increase in the number of myotomes and development of internal organs, which have been described in Table 2 and Fig. 7H to 7N. The embryo hatched in 24.15±01.00 pfh (26±1°C). *Marimuthu and* Haniffa (2007) have described the embryonic development of *C. striata* with more or less similar recordings.

2.2.8 Larval Development

Details of larval development up to yolk absorption stage are given in Table 4 and Fig. 8 (Rawat *et al*., 2020). Newly hatched larva is translucent, dull brown in colour and has well defined yolk. The head and yolk together looked like a bulb and anal portion has a distinct transparent caudal fin-fold having approximately 38 myotomes. It measured 3.14mm in total length (1.61mm thoracic length + 1.42mm anal length) with yolk length and yolk width of 1.20mm and 0.80mm respectively. Marumuthu and Haniffa (2007), however recorded size of newly hatched larvae as 3.4mm, which is slightly larger. The hatchling moves with jerks up and down but unable to swim (Fig. 8A). The heart was functional but flow of blood was not visible as being transparent at this stage. In subsequent developments, the hatchling increased both in head and anal lengths (Fig. 8B & 8C). The rudimentary eyes and mouth became prominent at 8.00±0.20 phh (32.15±0.20 pfh) (Fig. 8D). At 25.15±0.20 phh (49.30±0.20pfh), eyes are formed having dia of 0.35mm and distinct mouth appeared (Fig. 8E). The larva started vigorous swimming movement at 52.15±0.30 phh (76.30±0.30 pfh) with well developed mouth and pectoral fins. At this stage, yolk-sac is reduced to a very small size (Fig. 8F). Yolk-sac is totally consumed at 77.15±0.30 phh (101.30±0.30 pfh) when hatchling attained a total length of 6.20mm and freely move and voraciously consume live feed (Fig. 8G). These developments were found more or less similar as reported by Marimuthu and Haniffa (2007) in case of *C. striata* except that the length of yolk-sac absorbed larvae was recorded lesser (5.8mm) than in the present case. However, a measurable difference was observed in comparison to *C. punctatus* in terms of total length and anal length of yolk-sac absorbed larvae which were comparatively larger 6.56 mm and 3.99 mm respectively in the instant case in comparison to 5.80 mm and 2.90 mm in *C. punctatus* as reported by Haniffa *et al.* (2003).

Table 3: Details of embryonic development of striped murrel at 26±1 °C (n=4)

Stage	Post fertilization duration	Fig. No.	Developmental details
Fertilized egg	-	-	Eggs are round, transluscent, non-adhesive, floating, and pale yellow in colour having diameter of 1.50±0.05 mm. Diameter of oil globule and yolk-sac was respectively 1.1±0.1 mm and 1.25±0.1 mm.
2-cell stage	20:00±04.00 m	-	First cleavage
8-cell stage	30.00±05.00 m	2 (A)	Second cleavage
16-cell stage	45.00±05.00 m	2 (B)	Third cleavage
64-cell stage	60.00±10.00 m	-	Fourth cleavage
Morula stage	01.45±00.15 h	2 (C)	Blastoderm cell walls become undifferentiated;
Blastula stage	03.50±00.15 h	2 (D)	Embryonic shield formed and invaded around half of the yolk.
Gastrula stage	05.45±00.15 h	2 (E)	Gastrulation converts the blastoderm into a two-layered structure, with an outer epiblast and inner hypoblast. Head and tail region not differentiated
Early neurula	06.30±00.30 h	2 (F)	Cephalic region distinct with fore brain and narrow tail region. Formation of germ ring is seen.
10 myotome	11.30±00.30 h	2 (G)	Cephalic region further broadened & primitive brain visible, 10 myotomes visible, tail extends up to half of the yolk periphery, heart rudimentary.
15 myotome	12.15±00.15 h	2 (H)	Length of embryo further extends, around 15 myotomes visible, heart formed, eye lens formed in the rudimentary eyes kupfers vesicle.
22 myotome	13.45±00.30 h	2 (I)	Length of embryo further extends, around 22 myotomes visible, heart well developed and commenced blood circulation, eyes rudimentary,
25 myotome	14.15±00.15 h	2 (J)	Length of embryo further extends and cover around 80% yolk surface, heart well developed and commenced blood circulation but pulse rate very slow, around 25 myotomes visible, melanopores visible over the yolk-sac, tail narrow
27 myotome	18.30±00.15 h	2 (K)	Length of embryo further extends and cover around 85% yolk surface, cephalic and tail regions further widened, heart beat rhythmic and fast, melanophores increased in number
35 myotomes	19.00±00.15 h	2 (L)	Embryo covers entire yolk, cephalic region further enlarged, tail started detaching from yolk,

Stage	Post fertilization duration	Fig. No.	Developmental details
			melanophores increased further in number, around 35 myotomes visible,
Pre-hatched embryo	22.15±00.30 h	2 (M)	Tail detached from posterior side and moves at interval, myotomes more than 35, heart showed flow of red blood, eye precursor visible, auditory vesicle formed, twitching movement initiated, melanophores appeared on over the body.
Hatching embryo	24.15±01.00 h	2 (N)	Embryo coming out of egg shell with more vigorous twitching movement, myotomes more than 38, melanophore denser over the body, eye rudimentary
m=minutes, h=hour			

Table 4: Development of hatchling to spawn stage of striped murrel at 26±1 °C (n=4)

Phh time	Pfh time	Fig.No.	Developmental details
0.0	24.15±0.30 h	3 (A)	Hatchling dull brown in colour, yolk-sac well defined, caudal finfold transparent, myotomes around 38, verage total length 3.14mm, anal length 1.42mm, yolk length 1.20mm, yolk width 0.80mm, length of eye optic 0.34mm, eye optic distinct but eyes are not developed, mouth not formed. The larvae move up and down with jerks but could not swim.
5.15±0.30 h	29.30±0.30 h	3 (B)	Average total length 3.40mm, anal length 1.38mm, yolk length 1.44mm, yolk width 1.20mm, melanophores increased in number, 8.00±0.30 h 32.15±0.30 h 3 (C) Average total length 4.08mm, anal length 2.06mm, thorax length 2.02mm, yolk length 1.42mm, yolk width 1.08mm, eyes rudimentary
8.45±0.20 h	33.00±0.20 h	3 (D)	Rudimentary eyes and mouth prominent, average total length 4.08mm, anal length 1.85mm, yolk length 1.47mm, yolk width 0.80mm
25.15±0.20 h	49.30±0.20 h	3 (E)	Eyes and mouth formed, Eye dia 0.35mm, average total length 4.03mm, anal length 1.95mm, thorax length 1.78mm, yolk length 1.43mm, yolk width 0.79mm
52.15±0.30 h	76.30±0.30 h	3 (F)	Average total length 4.96mm, anal length 2.99mm, thorax length 2.0mm, eye dia 0.35mm, head width 1.0mm, yolk-sac reduced to 1.29mm, pectoral fins developed and larvae move vigorously for swimming, mouth gape well formed
77.15±0.30 h	101.30±0.30 h	3 (G)	Average total length 6.20mm, anal length 3.99mm, thorax length 2.59mm; eyes increased in size with eye dia 0.42mm and eye width 0.40mm; head width 1.29mm, mouth cleft 0.53mm, vestigial yolk-sac present, melanophores dense, pectoral & caudal fins developed

Phh time, Post hatch hour; Pfh time, Post fertilization hour

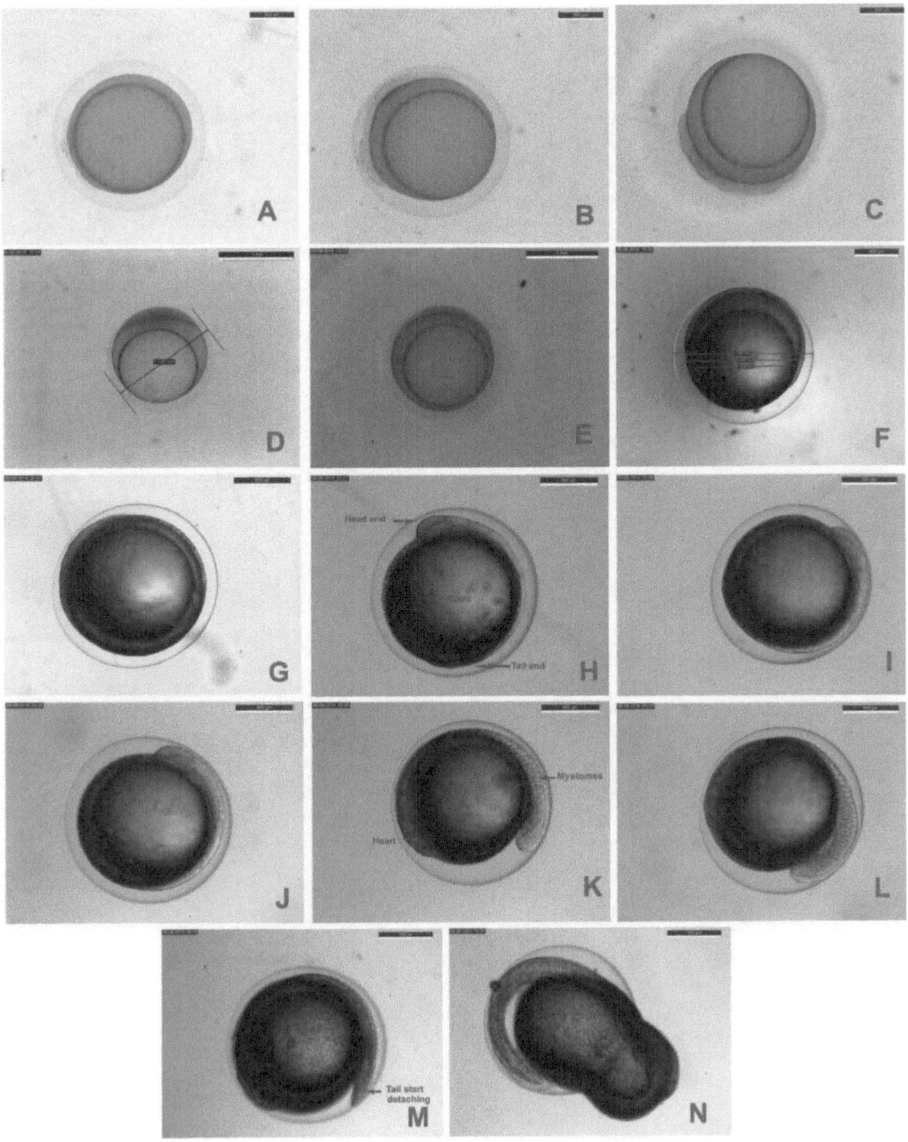

Fig. 7: Developmental stages of egg of striped murrel till hatching

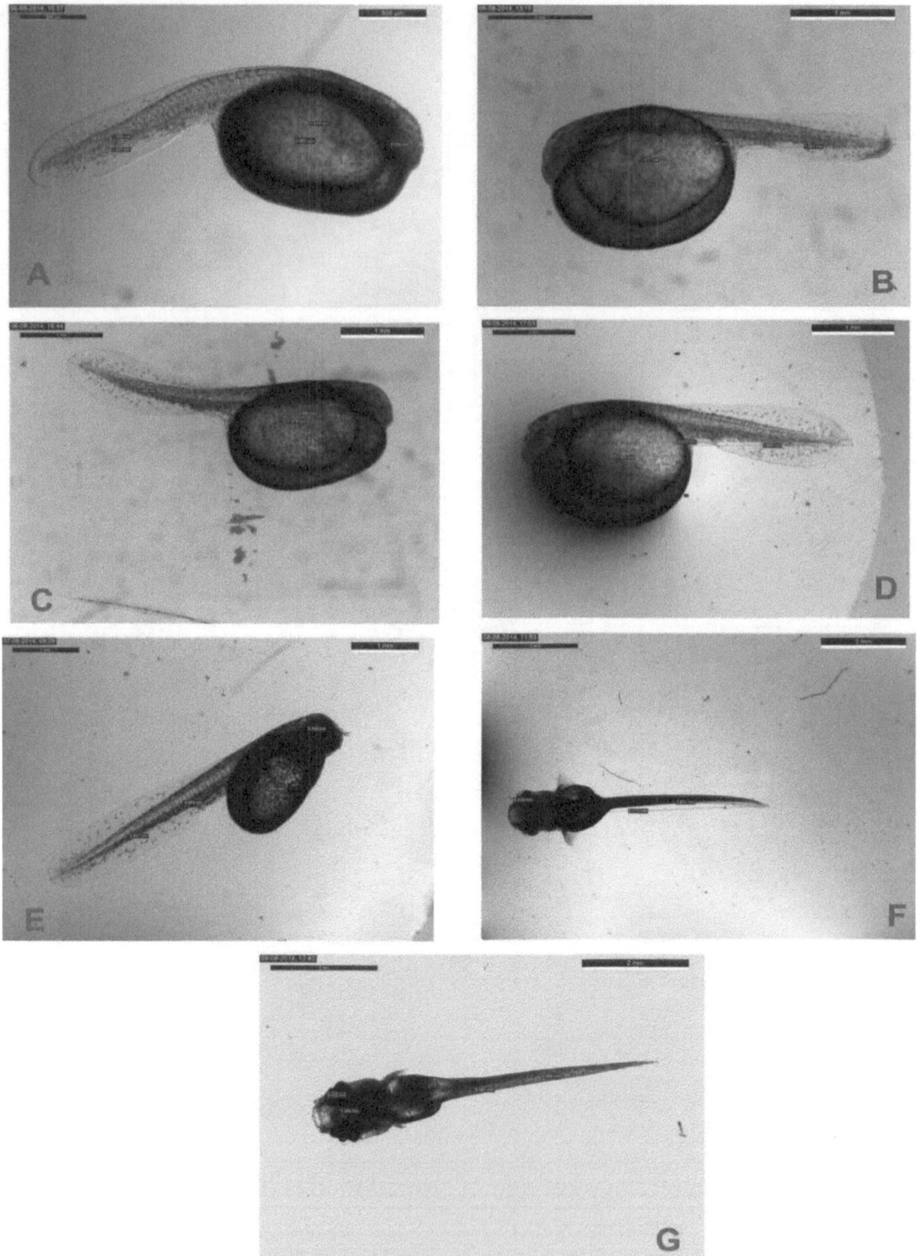

Fig. 8: Developmental stages of hatchling till yolk-sac absorption of striped murrel

3

Spawning in Wild

3.1 Spawning Season

Striped murrel breeds in stagnant waters and commence breeding in ditches, ponds, tanks, flooded paddy fields, wetlands and dead end of rivers; a little prior to or with the onset of monsoons (Bhattacharya 1946; Chacko 1947). Striped murrel is a batch spawner species and hence breeds several times in a year. According to Courtenay *et al.* (2004), spawning occurs in this species all through the year. In northern states of India, the breeding season of this species coincides with south-west monsoon and extends from June to September with peak in mid of June to mid of August (Qasim and Qayyum, 1961). However, the species has been reported to breed during November to January in southern India (Alikunhi, 1953).

3.2 Spawning Behaviour

Courtship in both male and female brooders occurs on attaining maturity and favourable climatic conditions. Pairing occur between a single male and female for spawning (Parmeshwaran and Murugesan, 1976a). The courting pair moves from one place to another for selection of suitable spawning site. A sheltered and shallow weedy margin of water bodies, unaffected with wind, is selected for egg laying (Khan, 1924). Female laid the eggs in batches at one place, which are immediately fertilized by the embracing male. The whole egg-mass referred sometime as nest thus remain at one point and surrounding with aquatic weeds and leaves of terrestrial plants (Willey, 1909; Pethiyagoda, 1991) (Fig. 9). The eggs hatch within 54 hours at 16-26°C and 30 hours at 28-33°C (Parameswaran and Murugesan, 1976b). The hatchling having long yolk-sac remain sheltered under the aquatic weeds at this time. The yolk-sac is completely absorbed on 4th post hatching day (phd) and the larvae become free floating and move freely in a shoal in water. The eggs and larvae are guarded by the parents. Generally male remain dominant in parenting, while female also move nearby and guard them, whenever male move away in chasing the attackers (*Pethiyagoda,1991*).The parental care is continued until the offspring's reach the total length (TL) of about 10 cm length (Breder and Rosen, 1966). The typical brood size of striped murrel is generally large initially, but with passing of time of about 15-20 days, only about 500-1000 young ones can be seen in a shoal. The reduction in number of larvae with passing of time is due to two factors, the filial & intracohort cannibalism (described in Chapter 5) and natural mortality.

Fig. 9: Nest of striped murrel containing egg mass in pond

3.3 Procurement of Seed from Wild Sources and Weaning on Artificial Diets

By and large, the farming of striped murrel in most of the countries solely rely on procurement of seed from the wild resources as hatchery seed availability is negligible. As wild seed having the inherent behavior of cannibalism, they are cultured under extensive farming system in ponds having greater prey population for feeding by the striped murrel. The production from such farming, however remain uncertain and solely depend on the availability of prey throughout the culture period. However in recent, the wild seed is weaned to accept formulated feed and therefore cultured under monoculture semi-intensive and intensive farming systems in many countries with higher rate of return. Protocol of rearing fry procured from wild resources to fingerling size on formulated pellet diet is given in chapter 6 and 7 of this book.

4

Captive Spawning

In order to get assured supply of quality seed weaned on formulated diet, the captive spawning is the right choice for successful farming of striped murrel. Captive spawning involves several activities such as broodstock development, induced breeding, hatching of eggs, larval rearing, etc. These activities are given in detail in the following sections.

4.1 Broodstock Development

Brood stock development is an integral part of captive breeding programme in fish and the success of breeding largely depend on it. It involves rearing of healthy brooders of both sexes in captive conditions to meet the targeted production of seed. In striped murrel, maturation of gonad begins in the first year of their life. However, brooders of 2^+ year age or above are better and recommended for captive breeding (Pati *et al.*, 2004). A size of 400g to 1500g is considered ideal (Fig. 10). Brooders may be procured from wild sources or reared on-farm from farm-raised fingerling. Brooders from wild stocks are to be procured at least 2-3 months prior to spawning season so that they acclimatize to the farm conditions. It is recommended to give them a bath of 50 ppm formalin for about 8 hours under vigorous aeration before stocking in the tank/pond. Formalin treatment removes external pathogens from the fish body. Brooders raised at the same farm do not require formalin treatment, if they are free from any wound.

The brooders collected from wild to be kept in separate tank/pond to avoid cannibalism, if they have too much difference in size among them. It should be noted that this species can consume a prey size ¾ in length than to the predator and accordingly size graded and separated. As all males do not attain maturity simultaneously along with females, it is better to maintain higher sex ratio of male brooders in comparison to female to get sufficient number of mature male brooders at the time of spawning. Cement tanks are better for keeping brooders collected from wild stock due to easy facility for frequent water replenishment and collection with small net and less manpower.

Fig. 10: Brooders of striped murrel

A stocking density of 2kg/m^3 is considered reasonable in cement tanks with fortnightly/monthly partial change of water (Pati *et al.*, 2004). In case of rearing brooders in a pond, a maximum stocking density of 3000-3500 kg/ha is reasonable (Mollah *et al.*, 2009).

Brooders reared in pond, may also be shifted to cement tanks few days prior to undertaking spawning for easy operation. About one-third surface area of the tank may be provided with water hyacinth (*Eicchornia crassipes*) or water lettuce (*Pistia* spp.) to provide a shade and a natural environment to brooders during non-breeding season but should be removed on the onset of spawning season, otherwise natural spawning may occur.

4.2 Brooder Diets

Feed is one of the most important components of healthy brood development. The protein requirement of adult striped murrel has been estimated in the range 49.2-55.6% (Mohanty and Samantaray, 1996), which is sufficient to meet the nutritional requirement for brooders also. Hua *et al.* (2019) in a recent study has estimated crude protein and digestible protein requirement of this species to be 52.1% DM and 47.8% DM, respectively. However, the demand for lipid increases with onset of maturity as lipid has an important role in the growth and maturation of oocyte. Higher lipid diet significantly increase production of oocytes per female and reduce the number of broodstock needed to meet egg production goals. It also increase egg quality, fertilization success and egg hatching success in fish due to presence of dietary EPA (Eicosapentenoic acid) and DHA (Docosahexaenoic acid) (Bobe and Labbe, 2010; Lupatsch *et al.*, 2003; Tyler, 1994; Izquierdo *et al.*, 2001). In order to meet the level of lipids in the diet of striped murrel brooders, a dietary levels 180g.kg^{-1} has been recommended (Ghaedi, *et al.*, 2016; Marshall *et al.*, 1999). However, lipid level above 190g.kg^{-1} in diets did not spare dietary proteins from catabolism in striped murrel and should not be used (Hua *et al.*, 2019).

Striped murrels collected from wild do not accept pellet diet and hence fed with their natural diet such as small fish, cut-fish pieces, shrimps, mollusk etc. A diet of 5% bw is sufficient to satisfy their feed demand. As this species do not over feed at a time, it is better to split feed demand into two equals for giving during morning and evening. Care should be taken that the size of the food to fit well in the mouth of the fish. In case of having pond reared brooders previously weaned on pellet diet, the feed has to be improved with higher levels of protein and lipids as suggested above i.e. protein level of about 45-50% and lipid level of 18% at least two month prior to peak spawning season.

4.3 Water Quality

The striped murrel grows well in normal freshwater water conditions as well as in low saline waters. It can tolerate comparatively wider changes in water quality parameters than other species. The ideal water quality for striped murrel rearing and spawning is temperature 27-33°C, Sechhi disc transparency 25-45 cm, pH 7.5-8.5, total alkalinity 100-150 mgL^{-1}, free CO_2 3-6 mgL^{-1}, dissolved oxygen 5-7 mgL^{-1}, dissolved inorganic phosphorus 0.1-0.2 mgL^{-1}, nitrate-N <0.03 mgL^{-1}, nitrite-N < 0.01 mgL^{-1} and ammonia- N <0.01 mgL^{-1}.

4.4 Selection of Brooders

The brooders are selected based on the secondary sexual characteristic as defined in Para 2.1 (chapter-2) (Figure 11 and 12). However, full maturity of a female could be better to assess by inserting canula in the ovary and drawing few ova. If the drawn sample show good amount of fully yolked-oocytes under the microscope, full maturity of female is established. Oocyte diameter between 1.0 to 1.3 mm is considered suitable for spawning (*Narejo et al.*, 2015). However, assessment of male maturity in murrel is a serious problem in captive breeding as secondary sexual characters are not very specific to determine full maturity as no milt is drawn on pressing the belly like carps and other species. Further, the milt is transparent and resembles similar to alimentary canal exude. Hence, males are generally selected based on the presence of distinct genital papilla and darker skin colour, which on times do not provide a conclusive proof (Hossain *et al.*, 2008; So, 2012; Irmawati, 2019). As determination of maturity takes considerably long time in murrel, it is recommended to anesthetize the brooders with either MSS-222 (25-50 mg/L) or clove oil (40-60 mg/L). A pair of brooder comprises of a ratio of 1:1 (female:male), as only one male cohabit with a female in striped murrel. However, in case of doubt, a sex ratio of 1:2 (female: male) ratio may also opt to off shoot chances of poor maturity in a single male. The brooders are transferred to hatchery at least 6-8 hours prior to induced breeding to acclimatize them in the new environment.

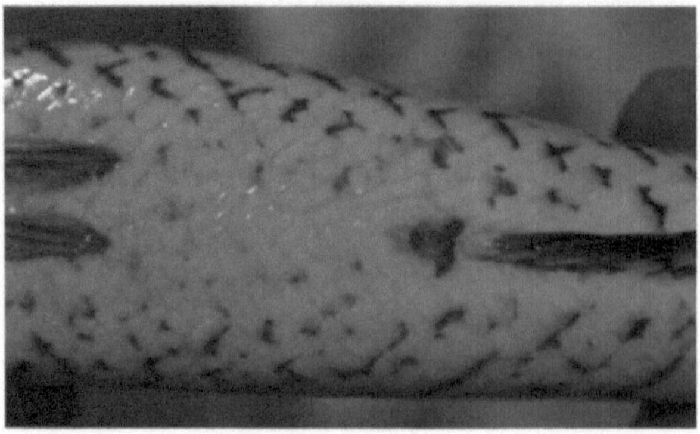

Fig. 11: A female with bulging belly, swollen vent and no space between vent and anal fin origin

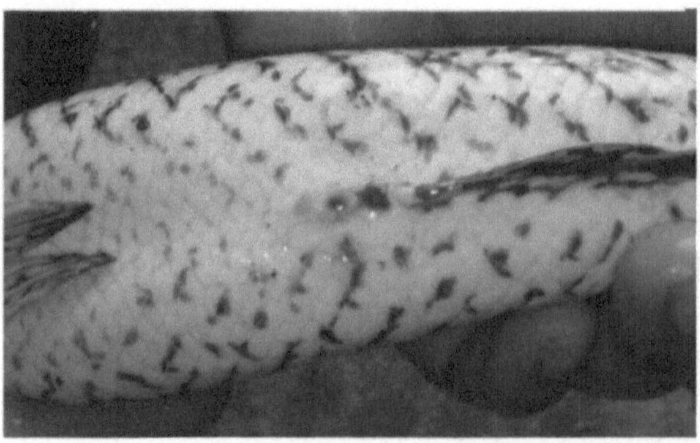

Fig; 12: A male with slender abdomen, un-swollen vent and distinct space between vent and anal fin origin

4.5 Induced Breeding

Induced breeding of striped murrel has been attempted in India, Bangladesh, Cambodia, Malaysia, Vietnam etc. (Yaakob and Ali, 1992; Hossain *et al.*, 2008; Bilal *et al.*, 2012; Dayal *et al.*, 2013; Rahman and Awal, 2016; Nen *et al.*, 2018; Rawat *et al.*, 2020) using natural and synthetic hormones. Various steps of induced breeding are discussed in the subsequent para.

4.5.1 Spawning Tanks

So far no specific hatchery design has been developed for spawning of striped murrel. However, separate facilities are required for acclimatization of brooders, spawning, incubation of eggs and larval rearing.

Spawning can be successfully undertaken in hapa fixed in the pond (Marimuthu *et al.*, 2007; Roy *et al.*, 2016), FRP (Fibre-reinforce plastic) tanks (Pati *et al.*, 2004; So, *et al.*, 2012; Rawat *et al.*, 2020), cement tanks (So, *et al.*, 2012) and earthen pits (Dayal *et al.*, 2013) (Figure 14, 15 16). The size of spawning enclosures may have surface area of 1-4 m^2. A water depth of 0.3 to 1.0m is sufficient as the species spawn in shallow area as in wild conditions. All the spawning enclosures are provided with macro-vegetation to cover about 25-30 percent surface area to create natural environment and space for laying eggs among macro-vegetation as in case of wild spawning. Suitable macro-vegetation commonly used is water hyacinth (Eicchornia crassipes), hydrilla or water thyme (*Hydrilla verticellata*) and water lettuce (*Pistia* spp). A single hapa/tank/pitmay be used for keeping 1-3 pairs of brooders at stocking density of one pair per square meter area.

Fig. 13: Shallow depth FRP tanks for indoor spawning

Fig; 14: Single breeding hapa fixed in pond

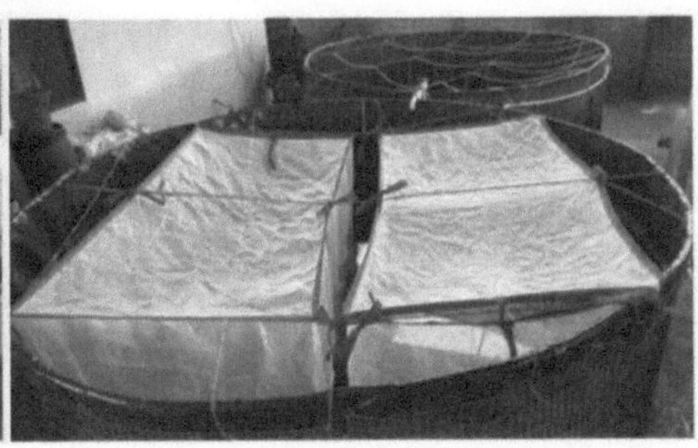

Fig.15: Breeding hapas fixed in indoor lined tank

Fig. 16: Selected two pairs of brooders

4.5.2 Inducing Agents

Various inducing agents such as carp pituitary gland extract (PGE) and commercial formulations; Human chorionic gonadotropin (HCG), Luteinizing hormone releasing hormone anologue (LHRHa) and salmon gonadotropin releasing hormone anologue (sGnRHa), have been used to spawn striped murrel (Francis *et al.*, 2000). CPE has been used in low to very high dose rates to breed this species with varying success rates (Alikunhi, 1953; Parameswaran and Murugesan, 1976b; Haniffa *et al.*, 2000; Hossain *et al.*, 2008; Roy *et al.*, 2016; ICAR-CIFA, 2016). HCG has been recommended for spawning of this species by Haniffa (2000), Bilal *et al.* (2012), So *et al.* (2012), ICAR-CIFA (2016); whereas LHRHa by Haniffa *et al.* (2000). However, performance of sGnRHa (Ovaprim, Ovatide, Ovupin, WovaFH, GonoproFH) has been found better in terms of its use, easy availability, cost effectiveness and success rates (Haniffa *et al.*, 1996; Haniffa *et al.*, 2000 & 2001; Marimuthu *et al.*, 2007; Pati *et al.*, 2004; Dayal *et al.*, 2013; Rahman and Awal, 2016; Roy *et al.*, 2016; Yulintine *et al.*, 2017; Rawat *et al.*, 2020). Selvaraj and Frances (2007) have recommended HCG implantation for faster induced maturation in both female and male striped murrel.

4.5.3 Doses for Inducing Agents

The dose rates of various inducing agents for female and male fishes tried by different workers for induced spawning of striped murrel is summarized in Table 5.

Table 5: Inducing agents and their dose rates used by various researchers for induced breeding of striped murrel

S.No.	Inducing Agent	Female	Male	Reference
1.	Pituitary extract	I dose:2-20mg/kg II dose:5-380mg/kg	I dose:2-20mg/kg II dose:5-250mg/kg	Parameswaran and Murugesan, 1976b
		I dose: 10mg/kg II dose: 100-150mg/kg	Not quoted	Haniffa *et al.*, 2007
		I dose: 40-60mg/kg II dose:80-90mg/kg	40-60mg/kg	Hossain *et al.*, 2008
		I dose: 2mg/kg II dose:6mg/kg	2mg/kg	Roy *et al.*, 2016
		30-40mg/kg	20-30mg/kg	ICAR-CIFA, 2016
2.	HCG	1000IU/kg	I dose:500IU/kg II dose:1000IU/kg III dose:2000IU/kg	So *et al.*, 2012
		4000-5000IU/kg	Not quoted	Haniffa *et al.*, 2007
		2000IU/kg	1500IU/kg	ICAR-CIFA, 2016
		4000-6000IU/kg	4000-6000IU/kg	Bilal *et al.*, 2012
3.	LHRHa	40-60μg + Pimozide 5 mg/kg	Not quoted	Haniffa *et al.*, 2007
4.	sGnRHa	0.5ml/kg (Ovaprim)	Not quoted	Yulintine *et al.*, 2017
		0.5ml/kg (Ovaprim)	0.2-0.4ml/kg	Dayal *et al.*, 2013
		0.5-0.7ml/kg (Ovaprim)	Not quoted	Haniffa *et al.*, 2007
		0.6ml/kg (Ovatide)	0.4ml/kg	ICAR-CIFA,2016
		0.4ml/kg (Ovatide)	Not quoted	Marimuttu *et al.*, 2000
		0.2-0.6ml/kg	Not quoted	Marimuttu *et al.*, 2007
		1.0ml/kg (WovaFH)	0.6ml/kg	Pati *et al.*, 2004
		0.5ml/kg+1.0ml/kg (Ovupin)	0.5ml/kg	Roy *et al.*, 2016
		1.0ml.kg (GonoproFH)	1.0ml/kg	Rawat *et al.*, 2020

In all above cases, the females are given higher dose rates of hormone than males excepting in case of So *et al.* (2012), Bilal *et al.* (2012) and Rawat *et al.* (2020). The higher rate of spawning success and shorter incubation period for spawning after hormone treatment indicated that higher doses to males are better for spawning of this species. As described earlier, male murrels have very small testis and assessment of full maturity is also difficult from external characters, application of higher dose rate of hormone to males is highly desirable. In fact males can even given more than one dose of hormone, one or two days in advance, before giving injection to female. This has been demonstrated by Nguyen and Long (2008) and Bui Minh Tam *et al.* (2008) in case of *C. micropeltes* and *C. marulius,* when they used higher doses of HCG in case of male fish than to female for successful spawnings. Bui Minh Tam *et al.* (2008) found HCG dose of 500 IU per kg female to be effective for induced spawning but males required higher doses of 3,500 IU per kg body weight in three split doses within a period of three days or 72 hours.

In most of the breeding trials given in Table 5, the successful spawning rate has not been clearly defined and different opinions have been expressed about the suitability of inducing agents and the breeding environment. Based on previous works and the personal experience of the authors, three criteria are suggested for spawning of this species. *Firstly*, careful assessment of full maturity of both the sexes; assessment of female maturity is easy but difficult for male. Hence, dedicated effort is needed for identification of male maturity as suggested earlier. *Secondary* dose rates of hormone, which needs to be applied to female and male fish. Females have been reported to successfully breed with low as well higher doses of PG, HCG, LHRHa and sGnRHa. But on many occasions, either no spawning occurred or female spawned with good quality unfertilized eggs that perished subsequently after spawning, when male fish is given low dose of hormone. The failure in spawning may happen due to failure of either one sex or both but in later case, the non participation of male fish in spawning would be the sole factor. Therefore, the males are advised to be given higher doses of hormone to trigger final maturation in testes as suggested by Bilal *et al.* (2012), So *et al.* (2012) and Rawat *et al.* (2020). So *et al* (2012) have observed higher spawning success when males are given higher dose of hormone within a span of 2-3 days in advance than to female fish. *Thirdly*, the spawning tanks should have shallow water depth and provided with submerged aquatic macro-vegetation sparsely-spread on water surface for allowing the brooders to laid eggs in a closed conformity in nest like manner as normally done by them in wild resources.

4.5.4 Maturity Advancement

Selvaraj and Frances (2012) and ICAR-CIFA (2016) have recommended HCG implantation for faster induced maturation in both the sexes of striped murrel. Pellets containing hormone are implanted in fish for sustained-release of gonadal hormone to advance and synchronize maturation of gonads. For making pellets, the required quantity of HCG hormone is mixed thoroughly in normal saline. Twenty parts of cholesterol and one part of binder (Gelatin: Gum Acacia in 1:1 ratio) are mixed well with the help of mortar and pestle.

HCG suspension is added with cholesterol mixed with binder. Ethyl alcohol (50%) is added and continuously mixed to get a gel-like consistency. From this paste, HCG Hormone Pellets of 2 mm x 4 mm size (oval/rice- grain shaped) and weighing 15-20mg are hand-madein. The hormone concentration of the pellets is adjusted according to the requirements and it may range from 200-1000 IU of HCG. The pellet is inserted in the brood fish body in the month of February to get maturity in over 80-90% brood fish in May. The testosterone level was found highest in the month of May (24.5pg/ml) in implanted male fishes. Similarly, highest level of estradiol 17 ß level (62.0 pg/ml) was also observed in May in female fishes. The GnRHa Pellet is also made similarly except that the mixture is kept at 35°C for drying to get a gel-like consistency. The pellet is implanted intramuscularly on the dorsal side of the fish. Small incision is made on the skin and the „Sustained-release Hormone Pellet is pushed into the muscle manually. After proper implantation of pellet, antibiotic cream is applied all around the incision. It has been observed that the wound gets healed completely in a week's time and only a tiny black scar remains after two weeks. The oval/ rice-grain shaped pellets are easy to implant and also require very small incision in the fish muscle. Therefore, fish are less stressed and the chances of rejection are also minimized.

4.5.5 Multiple Spawning

Striped murrel is a batch spawner as stated earlier. The first record of second breeding of striped murrel under captive condition was stated by Parmeswaran and Murugesan (1976a), who has reported two cases of second breeding after a gap of 75 and 142 days without any hormonal manipulation. Pati *et al.* (2004) have also reported second breeding in striped murrel in indoor conditions and have recommended that second breeding is possible after a gap of 60 days.

4.5.6 Preparation of Hormone Solution

The hormone solution, to be injected in fish has to be prepared when inducing agents such as CPE, HCG and LHRAa are used, as PG is muscular in nature and other two are in powder forms. However, sGnRHa based synthetic commercial formulations are available in liquid form and hence directly injected in fish. For preparing the solution of CPE, HCG and LHRHa, the volume of hormone is decided based on the hormone dose, approximate weight of fish, and concentration of required hormone per ml plus 10 percent extra volumes for spillage and marginal errors. As a thumb rule, volume of hormone injected per site should not less than 0.1 and more than 2 ml for a fish depending on fish size. Large size fish can easily absorb higher volume in their muscles in comparison to small ones. Keeping a volume of 1.0 ml per kg body weight is ideal for moderate size brooders.

The injection of CPE is prepared based on the total weight of the fishes to be injected, recommended dose rate and the required volume of extract to be injected in fish. Total volume of extract will include the total volume of PG extract to be injected plus 10% extra volume to compensate pilferage. For example, if the recommended dose of hormone is 20 mg/kg then 2 mg/kg (10% of total PG) extra quantity of PG has to be taken. The volume of normal saline to be adjusted in such a way that all the fishes can be injected with a single concentration of solution, keeping in mind the minimum and maximum permissible limits of injection volume as stated in the previous section. If this is not possible then concentrated solution may be prepared initially to provide suitable concentration for bigger fishes and the left over solution is then diluted to meet the requirement for small-size fishes.

4.5.7 Methods of Injecting Hormone

Two methods of giving injections to the fishes are in vogue, (A) Intramuscular and (B) Intra-peritoneal. Intramuscular injection is given in the thick musculature above the lateral line from base of dorsal fin to caudal peduncle area. In small fishes, it is given near the origin of dorsal fin so that the syringe needle easily penetrate through the small body scales and inject the hormone in thick musculature for easy absorption. However with bigger size fishes, the injection site may be shifted towards caudal part to offset problem of penetrating syringe needle through large body scales. Intra-peritoneal injection is however better option for both small and large size fishes as very little portion of the syringe needle is required to be inserted in the body. It is given in the coelomic cavity at the base of pectoral fin, where scales are minute and syringe needle can easily penetrate the musculature in both small and large size fishes (Fig. 17). The syringe needle has to puncture only the thin musculature available at the base of pectoral fin without penetrating the internal organs. This can be easily felt as when needle is inserted in the musculature, some pressure is required but after piercing the musculature, the needle will move internally freely. Now the hormone has to be pumped at this point in the coelomic cavity.

Fig. 17: Intra-peritonial injection being given to male fish

For giving hormone injection to the brooders, both male and female brooders are collected from the acclimatization or conditioning tank. It is recommended to give them anesthesia, as suggested in Section 4.4 to avoid any stress to the brooders. The brooders are weighed and injection dose rate decided based on maturity status of the brooders. Brooders are laid over a thick cushion and given injection. In case the brooders are given anesthesia, they are kept in a tank containing freshwater with having provision for vigorous aeration till anesthetized brooders return to normal condition.

4.5.8 Spawning and Egg Laying

The hormone injected brooders are shifted to the spawning tanks and covered from top to avoid escape due to jumping. However in deeper tanks, where large free-board is available, there is no need to cover them. Bilal *et al.*, (2013) and Marimuthu *et al.* (2007) have described the courtship behavior and process of spawning in striped murrel. The hormone-induced fishes show breeding stimulus after four hours of giving injection. A pair is made between a single male and female. When two males are kept along with a single female, a pair between a single male and female is made and other male is rejected. Courtship is followed with elaborate mating behaviour. Spawning rituals commence in after 10 to 12 hours or more of giving (sGnRHa) hormone injection, when both male and female fish embraces each other with slow movement in a circular fashion and often apply abdominal jerks for release of gamets. When the eggs are about to release, both male and female move to bottom of the tank in embracing stage settle at the bottom and the female start releasing eggs, which are simultaneously fertilized by the embracing male. The eggs immediately move to surface of water and assemble at one point as a mass of eggs (Fig. 18). A good spawning is generally noticed within 17-24 hours of giving injection but sometime it may take longer time too. Eggs laid after 36 hours of giving hormone treatment are generally unfertilized and perish. Each pair makes a single egg mass of thousand of eggs in between the provided macro-vegetation (Fig. 18).

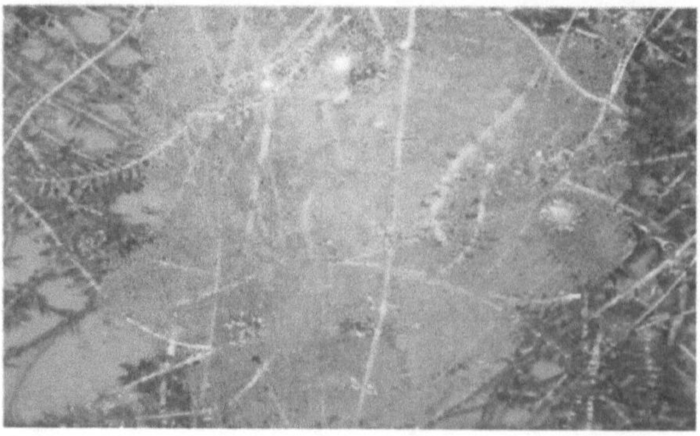

Figure 18: A large egg mass of fertilized eggs in FRP tank

4.6 Collection and Hatching of Eggs

The hapa/tank is to be checked for spawning of eggs after about 17 hours of giving final hormone injection. Once the spawning occurs, it is advisable to collect the eggs as early as possible as sometime brooders cannibalize them at the time of danger. If more than one pair is kept in a single spawning tank, the tank has to be checked for spawning after every 5-6 hours of collecting eggs of first pair. The eggs can be collected with plastic tumbler or glass beaker/tumbler by scooping (Fig. 19). The tank to be examined for spawning of all pairs till 48 hours and after that all brooders whether spawned or not, to be shifted for further rearing in tank/pond. Generally, a compact egg mass indicate successful spawning with fertilized eggs. Eggs found scattered over in small numbers in the tank indicate unsuccessful spawning and likely to have unfertilized eggs.

Fig.19: Eggs are being collected from spawning tanks

The eggs are floating, non-adhesive and having diameter of 1.20 to 1.40 mm (Fig. 20). Plastic tubs or small FRP/ cement tanks fitted with hapa can be used for hatching of eggs. The eggs spawned by more than one fish nearing the same time may be kept in same container but, if they are spawned at too different time, it is better to keep them in separate containers. Late spawning eggs are generally poor in fertilization rate and bad in quality and hence should not be mixed with those which have spawned within 24 hours. About 10000 eggs can be kept in 100 litre water container with water depth of about 10-12 inches. Providing aeration from an air- pump allows eggs to move from one place to another and also support water quality management and hence recommended. About 30% of tank water has to be replenished after a gap of 6-8 hours till hatching.

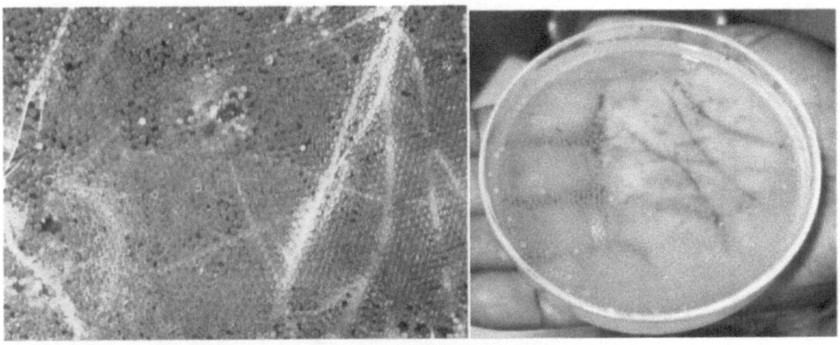

Fig. 20: Fertilized eggs in plastic mug and a petri dish

Fig. 21: Hatchling of striped murrel

Fertilization rate is generally high 60-95% but in some instances low as well to 28% (Parmeshwaran and Murugesan, 1976b; Roy *et al.*, 2016; Rawat *et al.*, 2020). Hatching rate depends on fertilization rate and may high up to over 90% when fertilization rate is high. Eggs hatch in 21-35 hours at temperature range of 24-32°C. Newly hatched larvae are ~3. 4 mm in length (Pati *et al.*, 2004., Thumronk *et al.*, 2011; Dayal *et al.*,2013; Rahman *et al.*, 2016; Roy *et al.*, 2016; Rawat *et al.*, 2020). The yolk-sac is completely absorbed in about 101±0.5 pfh with success rates of 50.76-93.18 percent (Rawat *et al.*, 2020). Thumronk *et al.* (2017) have reported suitability of saline water up to 12 ppt for hatching of eggs and up to 10 ppt for larval rearing, which open avenues of using saline water for hatching and larval rearing of this species, where freshwater is a scarcity.

5

Cannibalism and Mitigation

Predation and cannibalism are entirely two different processes. In predation, the animal used to consume other species then their own, whereas cannibalism is a practice of eating one's own kind. Cannibalism is a common behavior in animal kingdom and has been recorded in more than 1500 species. It is reported in an increasing number of fish species belonging to group/ families such as air-breathing catfishes (*Clariidae*), pikes (*Esocidae*), percids (*Percidae*), characids (*Characidae*), latids (*Latidae*), gadids (*Gadidae*) and in over 30 other families, including cyprinids (*Cyprinidae*) and salmonids (*Salmonidae*) (Smith & Reay, 1991; Hecht & Peinaar, 1993; Folkvord, 1991; Baras, 1998; Qin *et al.*, 2004). Cannibalism is a common ecological interaction and depends on genetics and fish behavior; mediated by appetite and metabolism and is promoted by size differences, and is more intense among fast growing larvae and young juveniles than among older fish (Hecht and Pienaar, 1993; Baras, 1998).

Cannibalism is a serious problem in the farming of striped murrel that starts from the early larval stage (Ng and Lim, 1990) and continues at all sizes including during parenting also. Murrels have large mouth gape; sharp jaw teeth for catching & grasping prey and can consume prey fish more than half to two-third of its length (Wee, 1981; Diana *et al.*, 1985; Ng and Lim, 1990 and Qin and Fast, 1996). Its aerodynamic body shape allows fast movement in water and catching the prey with ease. Striped murrel observes two types of cannibalism, Filial and Intracohort, which is discussed in detail in the subsequent sections.

5.1 Filial Cannibalism

Adult (Parenting) murrels observe a specific type of cannibalism; the filial cannibalism, in which parents eat their own offsprings. It is also recorded in at least seventeen different families of teleost fishes (Maniaca, 2002). Striped murrel observes parental care since laying of eggs to about 40-days old larvae, which is generally male-sex dominated. As parents always remain busy in parenting their offspring, they are not in a position to effectively prey upon their natural food; generally being, shrimp, fish, mollusk and insect larvae to satisfy their hunger. Under such conditions, parents consume their offspring to meet their energy requirement from time to time (Rohwer, 1978). This being the reason, the number of larvae in a parenting shoal of murrel reduces in numbers with aging of larvae and sometime all the larvae are consumed over a span of time. Filial cannibalism could be managed to a greater extent by providing supplementary feeding (Rohwer, 1978 and Sargent, 1992). Hoelzer (1992) showed in a field study that, if Cortez damselfish, *Stegastes rectifraenum* were fed with conspecific eggs, they cannibalized their brood less but still ate some of their own eggs. However, there is no such report in case of striped murrel.

5.2 Intracohert Cannibalism

Intracohert cannibalism (ICC) refers to eating of smaller larvae by larger ones in a population of their own. ICC is one of the biggest hindrances in the farming of many species including murrels (Hecht and Pienaar, 1993; Baras and Jobling, 2002; Baras, *et al.*, 2013) that cause extremely poor survivals during both larval rearing and grow- out phases. ICC in fish can start within a day after hatch, when only small portion of yolk-sac is absorbed and also between individuals of similar size, as observed in dorada (*Brycon moorei*) (Baras *et al.*, 2000). Larvae of longfin batfish (*Platax teira*) and Indian freshwater shark (*Wallago attu*) exhibit cannibalism behaviour at the age of 24 hph when major part of yolk-sac is not absorbed in the larvae (Leu *et al.*, 2018; Raizada *et al.*, 2015). In Striped catfish, *Pangasianodon hypophthalmus*, ICC is very severe and starts when larvae are only 40 hph old (Subagja *et al.*, 1999) and several sibling of uniform size in a group (up to 10 in number) can be found catching and damaging each others body parts (Baras *et al.*, 2010).

Responsible factors of ICC are both genetic and environmental. Major biotic factors for ICC are size difference, population density and availability of some weaker siblings in a tank whereas, abiotic factors include large number of parameters, such as food availability, food type, nutritional composition of food & its metabolism in body, temperature, photoperiod, shape & colour of internal walls of rearing tanks, etc (Baras, 1998; Smith and Reay, 1991; Britz and Pienaar, 1992; Folkvord and Ottera, 1993; Hecht & Pienaar, 1993; Baras *et al.*, 2000; Liao *et al.*, 2001; Baras and Jobling, 2002; Kestemont *et al.*, 2003; Szczepkowski, 2009; Kro´l and Zielin´ski, 2015).

Two types of ICC is dominant in fish; *Type-I*, the so called "early" type that occurs in the early larval phase and is independent of a diversity in fish sizes, where the victim is not completely ingested or consumed; and *Type-II*, a "later" form, associated with heterogeneous growth, when the victim (prey) is consumed as a whole (Hecht and Appelbaum, 1988; Smith and Reay, 1991; Baras and Jobling, 2002). A potential type-I cannibal mostly attacks its prey from the tail end, eats only a damaged part of the tail & abdomen and disgorges rest of the larva as it is unable to swallow it due to the lack of difference in mouth gape size of predator and body girth of prey. However in case of type II cannibalism, prey is attacked mostly from the head end or from the side but sometime also from the tail end and usually the prey is swallow completely (Baras and Jobling, 2002; Kestemont *et al.*, 2003). However because of a large predator-to-prey ratio in type II cannibalism, the mode of ingestion may be of little importance (Baras and Jobling, 2002). Type II cannibalism is determined by the mouth dimensions of a potential cannibal in relation to a potential prey, particularly to its body depth (Hecht and Appelbaum, 1988; Smith and Reay, 1991; Sogard and Olla, 1994).

The rate of Type-I and Type-II ICC varies from species to species. It ranges between 1.5 and 12.0 percent of the initial stock of European perch (*Perca fluviatilis*) larvae (Baras *et al.*, 2003; Babiak *et al.*, 2004; Kro′l and Zielin′ski, 2015) and from 10 to 17% in the stock of pikeperch (*Sander lucioperca*) larvae (Zake˛s′ and Demska- Zake˛s′, 1996; Zake˛s′, 2012; Kro′l and Zake˛s′, 2016) or even 40% in dorada (*Brycon moorei*) (Baras *et al.*, 2000). Type II ICC usually generates significantly higher losses and is mainly reported in a stock characterized by a large heterogeneity in size (Kro′l and Zielin′ski, 2015).

Qin and Fast (1996) have reported type-II cannibalism in striped murrel and similar is the observation of the authors as this species catch prey from either head or side, though on some occasions from tail end also and engulf it completely (Fig. 22). Small prey is consumed first than a bigger one, if both are present (Das *et al.,* 1998). In case the size of predator is too big to accommodate in gullet and stomach in one forego, the species takes several hours in engulfing the prey and on many occasions, the predator fish succumbs when unable to engulf the prey for a prolong time (authors unpublished communication) (Fig. 23). In order to assess the rate of ICC in striped murrel based on larval size difference, Qin and Fast (1996) developed a model to assess prey: predator TL ratio in a population of 10-120mm TL larval size, which cause cannibalism in murrel larvae. They reported 100% cannibalism in larvae when the TL ratio of smaller fish to larger was 0.35 in a 5 days trial. However, the rate of cannibalism decreased to 43% when the TL ratio of smaller fish to larger ones increased to 0.64, indicating that rate of cannibalism is higher in small size (TL) predators than the larger ones. It was also stated by them, that an 82 mm TL striped murrel could consume a 58 mm TL prey and a size of 65 mm TL were capable of consuming size of

68% smaller in TL, but fish 115 mm TL consume fish no longer than 50% their own body length. Thus the size of prey is determined based on the size of predator. A similar pattern was also observed in Atlantic cod, *Gadus morhua* (Folkvord, 1991). ICC triggers growth heterogeneity in larvae that depend for feeding on planktivorous food as digestible nutrients are often higher in fish than in plankton prey, which leads to the growth advantage of cannibals and, in effect, to the occurrence of type II cannibalism (Baras and Jobling, 2002; Kro'l and Zielin'ski, 2015). Cannibals can be easily identified by their gorgeous belly (Fig. 24).

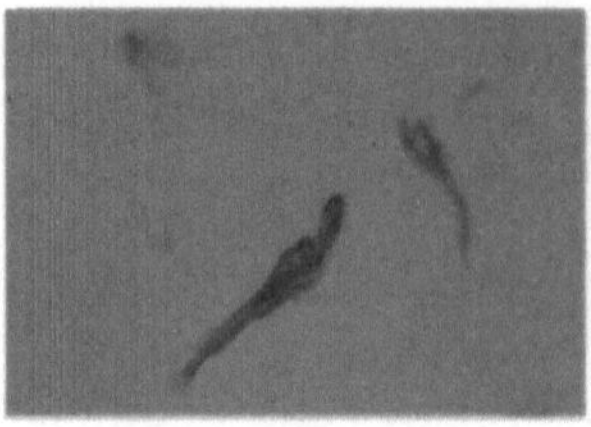

Fig. 22: Predation from tail end

Fig. 23: Predation from head

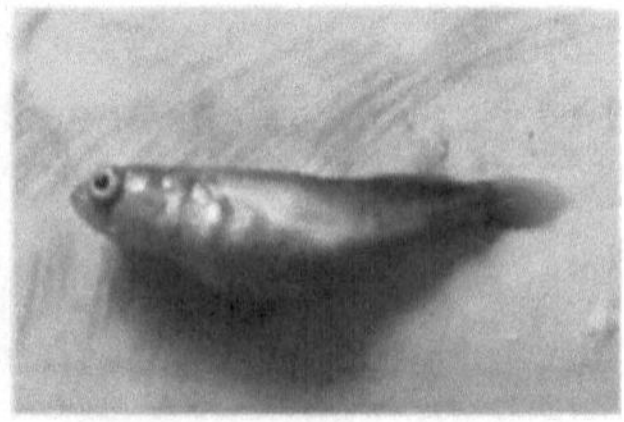

Fig. 24: Distended abdomen after complete engulfment of the prey

5.3 Measures to control Intracohert Cannibalism

Foremost measures to stop or minimize cannibalism in striped murrel, is to keep larval population uniform in size to the greatest possibility. The factors that could bring size homogeneity in larval populations are knowledge on larval age when it resort to exogenous feeding, optimum stocking density, type of suitable feed at different ages, satiation feeding, feeding frequency, live food supplements, temperature, photoperiod, etc. If these parameters are in check, chances of heterogeneity in size may be minimized and cannibalism would likely to be controlled to the maximum possible extent. Even if size difference is seen in a population, grading of larvae is another option to minimize ICC. Finally major measures that help in controlling ICC in striped murrel are application of efficient feeding strategy, use of feed supplements that minimize size heterogeneity and finally grading of larvae. These measures are discussed in detail as under.

5.3.1 Feeding Strategies

Age at which larvae initiate exogenous feeding, type of food requirement at different ages, quantity of feed, stocking density, feeding frequency, feed application methods are important parameters that can minimize or stop cannibalism during larval rearing. No cannibalism is seen in case of striped murrel, when larval population is uniform in size. It develops only when size heterogeneity occurs in a population. Delay in controlling size heterogeneity further aggravates cannibalism. Therefore our first aim should be the production of uniform size larval population in a rearing system. Several studies have demonstrated that although ICC aggression in fishes is difficult to stop, even after feeding the fish to satiation level but it can be reduced by increasing natural food availability (Fox, 1975; Qin and Fast, 1997a). But authors differ in their findings (unpublished data). According to them, ICC was not observed in hatchery produced larval population when 7 dph larvae of striped murrel were reared in hapa (made up of HDPE cloth, having 24 meshes per linear inch) fixed in an earthen pond, which was previously manured for sufficient production of plankton. A stocking density in these hapas was maintained @ 500larvae/m^3. The larvae were fed exclusively on the pond plankton entered through the meshes of the hapa and no other feed was applied from outside. The larvae reared for 21 days had almost homogenous size population with having minor heterogeneity of 24-28mm in TL and without production of any shooter in the population. It is important to note, that in above case the larvae were fed on BSN from the age of 48 hph to 7 dph in indoor conditions and had negligible heterogeneity at the time of stocking. Therefore, the primary cause of cannibalism in striped murrel is size heterogeneity in a population, which needs to be controlled since larva start feeding on exogenous food when only 48 phh only.

Once the size heterogeneity is developed in a population, cannibalism will certainly takes place. In that case the rate of cannibalism will depend on the food availability and size ratio between prey and the predator fish. Qin and Fast (1997a) developed a model to assess cannibalism in striped murrel larvae, based on application of feed and size difference ratio among prey and predator fish. According to them when two different size larval populations; larger fish of size 57.0±5.7mm TL and small of size 18.1±1.8mm TL in 1:2 ratio, were reared together without feed, 86% ICC was observed in captive conditions. However, cannibalism reduced to 60% and 35% when feeding ration was provided @ 5% and 15% of larval body weight in 6 days period. Rate of cannibalism also depend on quantity of feed application in a homogenous size population in this species. Cannibalism reduced from 12.23±1.94 to 3.31±1.90 percent with increase in feed application rate from 2 to 8 percent bw in a population of uniform size larvae (1.42±0.12g) (Kumari *et al.*, 2018).

Successful larval rearing of striped murrel therefore, depends more on the understanding of dietary requirements at the first-feeding and during subsequent growth phases. The rate of cannibalism increases in nutritionally poor environments as individuals turn to conspecifics as an additional food source. Cannibalism; however, does not-as once believed-occur only as a result of extreme food shortage or of artificial/unnatural conditions, but may also occur under natural conditions in a variety of species.

To reduce fish size heterogeneity and cannibalism in fish aquaculture, food abundance, feeding frequency and animal density are the most significant and common measures (Melard *et al.*, 1996; Baras *et al.*, 2011; Baras 2013). After murrel begin feeding on formulated diet, the progressive size variation as fish grow does not necessarily provoke cannibalism when an adequate amount of suitable food is available (Qin and Fast, 1998; personal observation of authors). Cannibalism in barramundi was substantially reduced or delayed by increase of food availability, which is consistent with the findings on other cannibalistic fish species such as European seabass (D. labrax) (Katavic *et al.*, 1989), striped murrel (C. striatus) (Qin and Fast, 1996) and Japanese flounder (Paralichths olivaceus) (Dou et., 2000). In a recent trial, the authors (unpublished data) have observed that when the larvae of length 62.64±0.16mm and weight 1.94±0.16g were weaned initially on live feed and subsequently on artificial pellet diet from 48 hph to 40 days post hatch (dph) age, cannibalism is drastically reduced with survivals of over 90% both in cement tanks and biofloc system of farming up to 8 months (the experiment is ongoing).

The authors therefore suggest the following feeding strategy for minimizing cannibalism in this species.

- The feeding to larvae should starts at 48 hph age, when it starts exogenous feeding.
- Feed to be acceptable to the fish, which changes with larval age.
- The larvae to feed with live feed of required size up to 25 days.
- The true digestion starts in this species at 25 dph, therefore inert diet should be avoided by this time.
- Always feed fish to satiation level so that it should not remain hungry any time.
- Always provide uniform distribution of feed so that all fish should get equal opportunity to feed.
- Feeding frequencies to be set as per the age of fish. High feeding frequency enables most fish to gain access to food and achieve homogenous growth (Katavic *et al.,* 1989; Baras and Jobling, 2002; Kestemont *et al.,* 2003).

5.3.2 Size Grading

Size difference among the larvae substantially increase the rate of ICC in a tank population and hence large size larvae or „shooters" are to be removed at the earliest to minimize cannibalism. Keeping population of both large and small size fish for a prolonged time, further enhances the rate of cannibalism. Grading has to be done with the help of graders, whereas mechanical graders are costly and beyond the scope of purchase for small farmers, manual graders as shown in Figure 25 and 26 may be locally get fabricated and used with good precision. Plastic basins are better as they allow making holes nearer and thus increase the efficiency of size grading. Graders of different hole size are required for making grading of different size larvae. The graders are kept over a tank with water and larvae are kept in the grader from where small size larvae pass through easily, leaving bigger one in the grader from where they are shifted to the other tank. Grading is labour intensive and not very intensive and may cause stress and injury to the larvae and hence should be used with utmost care (Hecht and Pienaar, 1993; Baras and Jobling, 2002; Hseu, 2002).

Fig. 25: Manual fish grader of plastic

Fig. 26: Fish size graders made of tin

5.3.3 Feed Supplements for Cannibalism Reduction

According to Changa *et al.* (2019), cannibalism is affected by serum cortisol, brain serotonin (5-hydroxytryptamine or 5-HT) levels and their interaction. Tryptophan (TRP), which is a precursor of serotonin (5-HT) and known to affect food intake and aggression in vertebrate, including fishes has been found to be effective in increasing the levels of 5-HT in the brain of groupers (*E. coioides*) and pikeperch (*Sander luciioperca*) and lower down cannibalism, although on the expanse of little bit lower downing growth within 10-15 days at 0.5-2.0% level in diet. However, these effects were higher in larger fish, presumably the dominant one, and in those which can consume a higher proportion of the feed offered (Hseu, 2003; Krol and Zyzislaw Zakes,

2016). Sangavi *et al.* (2019) studied the effect of L-tryptophan on striped murrel fry (0.19 ±0.001 g) in feed in 45 days trial. Tryptophan added diet (0.6%) has 84% survival, which was found significantly higher without affecting specific growth rate (1.49%). However, adverse affect of adding tryptophan have also been reported as increase in brain 5-HT concentration significantly reduces feed intake and hence growth (Pedro *et al.*, 1998; Ortega *et al.*, 2005).

Two percent addition of dry BL10 (micro alga) powder (*Aurantiochytrium* strain BL10) used to increase the DHA content did not benefit growth but could significantly reduce size discrepancies in orange-spotted grouper fingerling (*Epinephelus coioides*) within two weeks. BL10 supplements increased the proportions and contents of HUFAs in the brains of *E. coioides*, increased brain 5-HT levels, and reduced serum cortisol levels (5-HT). The serum cortisol level exhibited a significant negative correlation with brain HUFA content, particularly arachidonic acid (ArA). However, brain 5-HT has been found positively correlated with the ratio of n-3/n-6 fatty acids in the brain. More research is needed to evaluate effect of feed supplements in reduction of cannibalism in striped murrel.

6

Larval Rearing

The objective of larval rearing is to mass-produce high-quality and healthy juvenile fish. In larval rearing, the management of both rearing environment and feeding regime are the most important aspects. Improving larval rearing technique requires understanding of larval morphology & behaviour and requirements of live food, artificial diet and environmental conditions at different ages of larvae (Liao *et al.*, 2001). Striped murrel is predatory as well cannibalistic species in nature. Therefore understanding about its behavior with respect to feeds and feeding, stocking density with growing age is utmost important. It is a general thinking that fish hatchlings do not require exogenous food until yolk-sac attached to their body is exhausted. But this is untrue with many carnivorous fish species, particularly those observe cannibalism. Larvae of such fish species start feeding much before the yolk-sac absorption; rather delay in getting feed make them weak or force them to feed on the conspecifics. Therefore in order to understand larval behavior of this species in respect of feeds and feeding, knowledge on its ontogenic development is crucial particularly in respect of age at which mouth opens; mouth gape size, food choice, size of food particle at different age-groups and finally the age when gut digestion starts.

Yulintine *et al.* (2017) have given ontogenetic details of striped murrel larvae since hatching. According to him, mouth opens and exogenous feeding begin in this species when hatchling grows to a size 4.63$\pm$0.25mm on second say of hatching. On day-3 after hatching, the buccopharynx channeled with the anterior intestine through a short oesophagus and rather narrow lumen, with having prominent liver developed by this time. Two pyloric caeca appeared on day-25 hph, indicating the transition from larval to juvenile stage and acquisition of an adult type of digestion. This chapter provides strategy for larval rearing of striped murrel based on ontogenic studies and work carried out by various workers in this field including that of authors.

6.1 Larval Rearing Systems

Plastic tanks (capacity 60-100 litre), FRP tanks (capacity 200 litre to 500 litre) and small cement tanks are suitable for larval rearing of striped murrel under indoor conditions. Round tanks are always better than rectangular in facilitating better larval movement as well easy cleaning of tanks. The tank may be provided with both inlet and outlet for supply and draining water, however this can also be done manually by siphoning using flexible PVC pipe. Providing aeration from a portable air-blower keeps water quality better and hence recommended. Tanks may be provided with mild flow of water or otherwise, 40-50% of the tank water has to be changed at the time of cleaning the tanks. Rearing in outdoor conditions can be undertaken in hapa (fitted in pond) as well in open cement tanks. Outdoor rearing facility has to be covered from top by netting to avoid fish predation by birds as larvae are bright red in colour and also surface dweller.

6.2 Water Quality

Water free from any dirt/suspended particles has to be used in the larval rearing tanks. Freshwater canal and bore-well waters are generally good sources of water but to be checked for its suitability. The suitable water parameters are temperature 26-29°C, pH 7.5-8.5, alkalinity 100-150 mgL^{-1}, total hardness <250 mgL^{-1}, DO 6-8 mgL^{-1}, dissolved inorganic phosphorus 0.1-0.2 mgL^{-1}, nitrate <0.03 mgL^{-1}, nitrite <0.01 mgL^{-1}, unionized ammonia (NH_3) <0.01 mgL^{-1}. The water should also be free from any chemical pollution.

6.3 Larval Feeds

Larvae of striped murrel need divergent food items of different sizes at particular age/size (Qin & Fast, 1997b; Abol-Munafi *et al.*, 2004; Kumar *et al.*, 2008; Sarower *et al.*, 2010; Thumronk *et al.*, 2011; Bilal and Haniffa, 2016). The larvae are selective feeders and choice their prey depending on their mouth gape size and preference for the prey size increases with advancing of growth. This shift in diet is due to increase in spacing between the gill rackers (the filtration mechanism in gills for retaining food) with advancing of age. Smaller the larvae, the spacing between the gill rackers are less that retains smaller prey for feeding by the larvae. However with increase in the size of larvae, the spacing becomes wider in gill rackers and hence the smaller preys are filtered out from the operculum and of no use to the larvae. Therefore, size of the prey/feed given to the larvae should be retained at the gill racker site for consumption by the larvae.

Larval feeding is one of the important tasks of seed production of striped murrel for survival and faster growth together with reduction/ stopping of intracohort cannibalism, which is one of the greatest bottlenecks in larval rearing of this species.

Another cause of concern is larval inability to produce digestive enzymes up to the age of about 25 days and hence feeding with artificial diet is not possible before it. Though enzyme supplementation in diet may helps in early application of feed. Therefore in order to wean the larvae on inert diet, a systematic approach of providing live and inert feeds have to be resorted. Various studies on live and inert feeds carried out to minimize cannibalism in this species are given as under.

6.3.1 Live Feeds

Several studies on feeding striped murrel larvae using BSN (Fig. 27), Cladocerans (*Moina*) *Daphnia, Ceriodaphnia*), copepods (*Thermocyclops decipiens*), tubificids (*Tubifex* spp) (Fig. 28, 29), bloodworms (*Chironomus* spp) (Fig. 30), mosquito larvae and formulated feeds have been undertaken (Qin and Fast, 1997b; Abol-Munafi *et al.,* 2004; Kumar *et al.,* 2008; Mehraj-Uddin *et al*., 2009; War *et al*., 2011). The outcome of all the results revealed, Brine shrimp nauplii (BSN) are the best source of live feed to begin larval feeding of striped murrel and could be continued up to 30 days rearing, if other live feed is unable to be sourced. The reasons being the small size of BSN (length 0.251-0.348 mm and width 0.1256-0.1464 mm; Abol-Munafi *et al*., 2004) that suits to mouth gape of even 48-49 hph larvae and its easy digestibility in the gut, where they are auto-digested with enzymes available in the body of prey and hence the larvae gut devoid of producing digestive enzymes does not matter for its digestion. It is a matter of fact that all the past studies on feeding of striped murrel larvae had begun without determining at what age the functional mouth is developed in the larvae after hatching and feeding has been undertaken after the yolk- sac absorption.

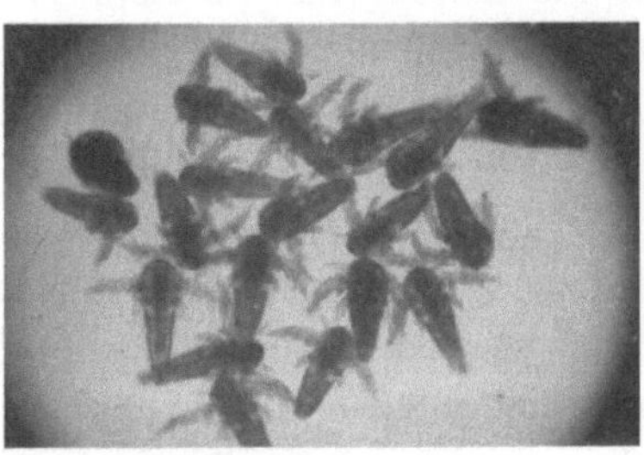

Fig. 27: BSN nauplii under microscope

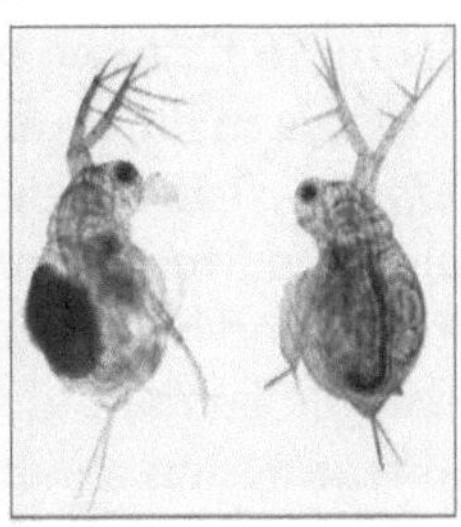

Fig. 28: Moina under microscope

Fig. 29: Tubificid worms

In striped murrel, the mouth is visible at 27$\pm$0.20 hph and become functional at 48-49 hph when larvae could also swim freely (Temperature 27-29°C) and start consuming exogenous food (Marimuttu and Haniffa, 2007; Rawat *et al.*, 2020). If exogenous food is not available at this time, the intracohert cannibalism (ICC) starts in striped murrel larvae, which is generally ignored during initial rearing phase. It is a matter of fact that once the Intra chohort cannibalism (ICC) is initiated, cannibalistic larvae gain growth advantage over non- cannibalistic siblings (Ribeiro and Qin, 2015) and if there is food restriction, the ICC become intense as the cannibal would persistently find conspecifics to prey on, further increasing size dispersal (Baras, 2013). Therefore in order to minimize/stop ICC and to enhance survival, the feeding to larvae with BSN has to initiate after 48 hph, so that the available food (BSN) could satisfy the appetite of larvae and minimize/ stop cannibalism (author's unpublished information).

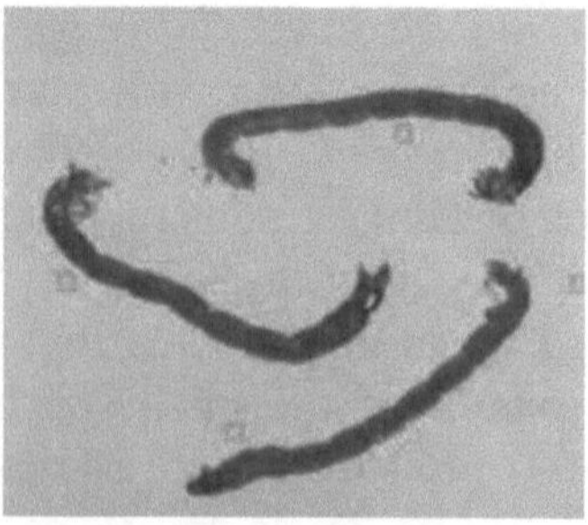

Fig. 30: Blood worms (*Chironomous larvae*)

Thumronk *et al.* (2011) have described that 62 hph striped murrel larvae accept *Moina* (length 0.605-1.355 mm and width 0.321-0.717 mm; Abol-Munafi *et al.*, 2004) at the mouth gape of size 324.30±144.60 µm. However, its average consumption was only 0.56 individual/ larva in the beginning, which is not sufficient to meet the overall nutritional demand of the larvae. Hence, *Moina* may not be considered ideal as first food for the larvae. The best survival has been obtained with BSN up to first two weeks of initial rearing, which is followed with *Moina* and Daphnia in the third and fourth weeks, though mixed cladocerans (*Moina*, *Ceriodaphnia*) and copepods (*Thermocyclops decipiens*) have also shown better results than Moina alone when fed to 16-day old larvae. Tubificids, mosquito larvae and bloodworms did not show promising results in comparison to other diets and the larvae did not sustain on formulated diet in the first two weeks of initial rearing but resulted in poor survivals during third and fourth weeks due to non secretion of digestive enzymes in the first two weeks and poor secretion in the gut during third and fourth weeks.

The better acceptability of BSN in contrast to *Moina* may be due to its smaller size in contrast to *Moina*. Therefore feeding with *Moina* has to be initiated at later stages, when larvae develop larger mouth gape for easy ingestion. Larvae of 7-10 day old are ideal to initiate feeding with *Moina* in phased manner with gradual replacement of BSN in about a weeks time (War *et al.*, 2011). Shifting the diet from BSN to plankton gives an advantage of better nutrition in terms of enzymes such as protease, carbohydrase, lipase together with EPA and DHA for meeting the nutritional requirements of the larvae. On the other hand, BSN are deficient in the level of essential polyunsaturated fatty acids, eicosapentaenoic acid (EPA) (20:5n-3) and docosahexaenoic acid (DHA, 22:6n-3), hence feeding them for a long time is not suggested unless BSN are enriched with both EPA and DHA. The methodology of BSN production and enrichment is given in Appendix-I. BSN can actively swim for 4 to 5 h in fresh water before sinking to the bottom and die thereby extending their availability for murrel larval consumption in live condition for only a small time due to its marine origin (Hoff and Snell, 1989; Qin *et al.*, 1997b). On the other hand, the survival period of freshwater plankton in freshwater is much longer than BSN, which increases chances of its predation by the larvae for a longer duration and chances of water getting fouled also minimized.

Therefore, it is obvious that BSN are the best source of live feed to begin larval feeding of striped murrel and feeding on BSN has to be initiated from 48 hph and may continued for 30 days, if *Moina* or similar crustacean plankton are not available. It is to note that only few larvae take up exogenous food in the beginning but with advancing of age their numbers are increased and almost all larvae start feeding within 60 hph age. Cladocerans (*Moina, Daphnia, Ceriodaphnia*) and copepods (*Thermocyclops decipiens*) to be given as live feed when the larvae are well weaned on BSN and have developed larger mouth gape for easy ingestion. Larvae of 7-10 day old are ideal to initiate feeding on cladocerans (*Moina* and *Ceriodaphnia*) in phased manner with gradual replacement with BSN in about a weeks time. However, 16 days old larvae may also be fed on copepods and mixed plankton comprised of cladocerans and copepods. Thus, the larvae to exclusively feed on live feed for first 25 days. Formulated feed enriched with digestive enzymes may, however be tried before 25 day post hatch day (dph).

6.3.2 Transitional Feeds

The diet used to shift larval feeding from live feed to formulated feed is herewith named as „Transitional Feed (TF). Initially, TF are supposed to be choice feed for the growing larvae but not dispensable for a long time due to ethical reasons, higher cost as well as water fouling problems. TF are given to larvae after the development of pyloric caeca, which enable production of digestive enzymes in the gut. In striped murrel two pyloric caeca appeared on 25dph, indicating the transition from larval to juvenile stage and acquisition of an adult type of digestion and a right stage to take up weaning on formulated diet (Yulintine *et al.*, 2017; authors personal communication). Fish and prawn meat are the choicest food for the striped murrel juveniles, which is readily accepted by them. Therefore weaning the larvae on formulated diet has to begin with mash of fresh or boiled fish/prawn meat for 3-5 days. Once the larvae starts feeding fish mash, grounded pellet feed (CP 45-50%, lipid 10%) has to be mixed in it with gradual replacement of 15-20% formulated diet every day so that the fish mash is totally replaced with formulated diet within 7-8 days and the larvae are totally weaned to feed on grounded formulated feed. The mash has to be in the form of thick paste, which can be compacted with hand, air dried and given a ball-shape. This way the larvae will be weaned on grounded formulated diet within 35 dph. The TF has to be given ad-libitum. Dayan *et al.*, (2016) have assessed better performance of pulverized shrimp meat when mixed with commercial diet (CP 36%) in comparison to freeze-dried krill.

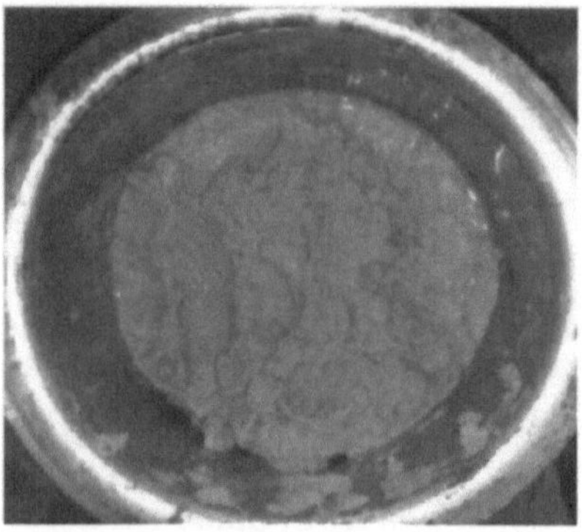

Fig. 31: Total fish mash

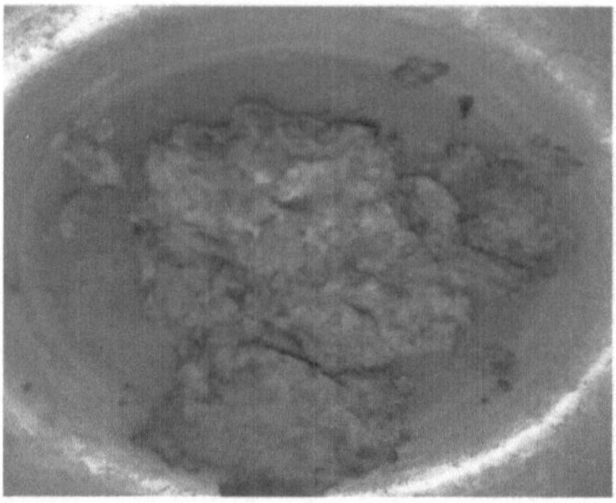

Fig. 32: Fish-mash mixed with grounded formulated diet

6.3.3 Formulated Feeds

Once the fry are weaned on transitional diet, feeding with floating pellet diet (Fig. 33) has to be taken up. The size of floating feed initially to be around 0.4 mm, which can be gradually replaced with 0.8 mm and 1.2 mm subsequently with increase in the size of larvae. Mixing of small and large pellet diet allows better feeding to both small and large size larvae. In case pellet of 0.4 mm and 0.8 mm are not available, the larger size pellet may be broken to small size in a mixer grinder or dough of grounded feed may be provide in a floating bowl in the tank till larvae attained size to eat the available size pellet feed. Giving feed in paste and powder forms creates problems of water fouling and increases bacterial load in the tank and makes cleaning cumbersome and hence not advisable for a long time.

Fig. 33: Larval pellet diet

Nutritional quality of formulated diet should match with the nutritional requirement of larvae. The protein requirement of striped murrel fry has been estimated to be around 50-55% on dry method (DM) basis (Mohanty and Samantaray, 1996), which is similar to protein requirement of giant murrel (*C. marulius*) fry i.e. 540–600 g.kg^{-1} (Raizada *et al.*, 2012). The optimum lipid requirement of striped murrel fry has been estimated 13 percent (Samantaray and Mohanty, 1997). The linkage between protein requirement and lipid levels has been estimated on *C. maculate* ♀ X *C. argus* ♂ hybrid for 8 weeks, which revealed that higher protein diets (46%, 52% and 57%) resulted in significantly higher weight gain in comparison to lower protein treatments (34% and 40%) and dietary protein significantly improved weight gain ($P < 0.05$) except the diet which had higher lipid (12%) and protein levels (57%) resulted in reduced growth (Aliyu-Paiko *et al.*, 2010). So a diet comprising of 45-50 percent crude protein and 10 percent lipid with 3.5 percent fiber is considered ideal to begin larval rearing.

6.4. Larval Rearing Methodology

Successful larval rearing of striped murrel therefore involves production of seed with optimum growth rate, higher survivals and total acceptance of commercial diet. Striped murrel larvae reach fingerling size in about 45 dph times for weaning on commercial diet before stocking in grow-out ponds. The feeding strategy suggested for raising fingerling has to be taken up in three phases of rearing, as given in the following sections.

6.4.1 Phase-I: Larval Rearing on Live Feeds

This rearing phase comprises of about 25 days in which larvae are exclusively fed on live feeds. Freshly hatched BSN are considered best live feed during this period (Abol-Munafi *et al.*, 2004; Bilal and Haniffa, 2016; Rawat *et al.*,

2020). BSN are very small in size and hence easily consumed by the larvae. Planktonic food such as *Moina*, Daphnia, *Ceriodaphnia*, tubificids, blood worms, etc. are bigger size food items and larvae unable to feed them during initial feeding. The larvae died when feed on artificial diet during this phase and hence should not be used during this rearing phase. As described earlier, feeding with BSN to start when larvae are 48 hph old even when their yolk-sac is not completely absorbed (Fig. 34). The Larvae at age looks-like a black slow moving mass in hatching tank as they move slowly in a compact shoal. If this is not done, the cannibalism starts by the dominant larvae and once the cannibalism starts, it becomes difficult to control at later stage. Hence, BSN feeding has to start in the hatching tanks itself. Once the yolk-sac of larvae is completely absorbed (generally on third day), the larvae have to be shifted to the designated larval tanks.

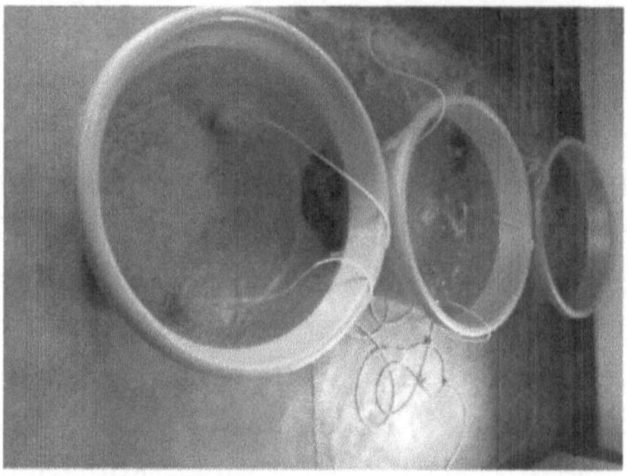

Fig. 34: Incubation of eggs, black colour due to development of eyes in the eggs

The larval tanks to be cleaned well and sun dried before use. They are filled with clean freshwater for undertaking larval rearing. The tanks are provided aeration from an air-blower. The larvae to be collected from hatching tank, washed with clean freshwater and treated with 25 ppm formalin for about 10-15 minutes before shifting in the tank as a prophylactic measure. The tanks to be stocked with yolk-sac absorbed larvae @ 4larvaeL^{-1}. The feeding has to begin with freshly hatched BSN to satiation. Though, BSN can be fed to larvae for one month rearing, its use has to be made on limited scale to cut down the cost of production and also due to deficient levels of EPA (20:5n-3) and DHA (22:6n-3). *Moina* is consumed by the striped murrel larvae at the age of 5-7 days (Abol-Munafi *et al.*, 2004) and with increase in larvae size, it also accept larger planktonic organism. The overall feeding strategy during this phase is to feed the larvae with freshly hatched BSN for a fortnight with gradual replacement of *Moina* after 7 days of yolk absorption (Fig. 35). This will minimize the operation cost on feed to greater extent. The feed to be given at least three times in a day: morning, afternoon and evening ad-libitum with BSN as BSN can survive in freshwater for about 5 hours only. To know how much live feed has to be given to larvae, it will depend on either the volume of water or the number of larvae in a tank. The thumb rule is to check the left over live feed before the subsequent feed. If there is no feed/negligible feed left at the time of subsequent feeding, it is an indication of low feed application in the previous feeding. If on the other hand, too much feed is observed as left over than it is an indication of higher feed application in the previous feed. After watching it, a worker may themselves adjust the required feed in a tank. Any larvae having size larger by one and half in length (shooter) to other larvae has to be removed and may be reared with similar size larvae in other tank. The tanks to be cleaned from all dirt every day during evening time together with change of around 30-40 percent tank water. The method for production of BSN and *Moina* and treatment of *Moina* & mixed-pond plankton are given in Appendix-I and II.

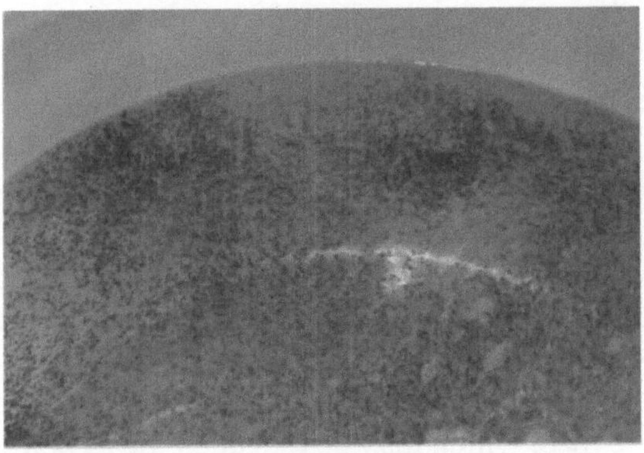

Fig. 35: Early larvae of striped murrel

6.4.2 Phase-II: Larval Rearing on Transitional diet (Weaning Phase)

With advancement of age, the larvae need larger food items as smaller are likely to be filtered out from gill rackers and will adversely affect the growth. Feeding large-sized live food items becomes difficult to feed and costly under mass seed production practices. Therefore larvae have to be fed on artificial diet to meet their nutritional requirement. Using artificial diet particularly commercial diet is simple to get and easy to give to the larvae with saving on production cost. However, weaning larvae to accept artificial diet is difficult and a tricky exercise. Therefore, second phase of larval rearing (weaning phase) is a bit complicated. The larval rearing may be continued in the same tank used in the first phase or may be undertaken in different tank(s) of same size or larger to maintain stocking density of 2 larvaeL^{-1} (Fig.36). Changing tank may avoid chances of parasitic and bacterial infections, which may occur due to prolong use of tanks. As stated earlier, the pyloric caeca develops in the larvae at an age of about 25-days, which reflect production of digestive enzymes in the larval gut. So at this point of time, larval diet has to be changed to non-living diets for which the larvae has to be weaned (trained) to totally accept cost-effective formulated diet.

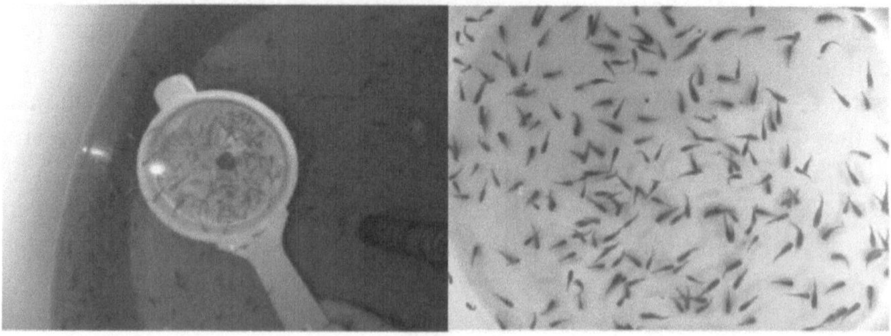

Fig. 36: Rearing of 15 day old larvae in FRP tank

The authors have found in their study that fine fish mash (raw/boiled boneless meat) easily attracts larvae to initiate feeding on it. Mash of small sized fish (SFS) is better due to their soft meat. Fish mash can be kept in small hanging bowl(s) one inch below the water surface in such a way that larvae can easily come in and leave the bowl after taking feed. Feeding with mash has to be initiated after 25 days of rearing on live feed. Since larvae remain most hungry during morning time due to night fasting, feeding with mash has to be initiated in the morning and continued till late afternoon to satiation. The evening feeding given to be of live feed, which may be of same type as being already in use. The larvae get trained to eat fish mash within 2-3 days.

Now, the larvae are to be gradually weaned on formulated feed with complete withdrawal of live feed and fish mash. The formulated feed used to be of same type & quality, which is likely to be used for subsequent rearing in fingerling (advanced size) or grow-out production tanks or ponds. The larvae do not accept formulated feed immediately and hence they have to be mimicked to accept fish mash which is gradually mixed with grounded formulated feed. Every day around 15% of the grounded formulated feed is added in fish mash with equivalent amount of withdrawal of fish mash in such a way that within next 7-8 days the whole feed comprised of formulated feed only. Thus changing feeding habit from live to formulated takes about 10 days time, when larvae reach age of around 35 dph. However, greater attention will be required towards cleaning of the larval tanks when mashed feed is given to the larvae. Mashed feed makes water more dirty and provide platform for pathogenic microbes and parasites and hence to be kept under control. Therefore during cleaning of the tanks, around 60-70 percent water has to be changed with clean water. Shooters, if seen in the tank, are to be removed either on daily or every alternate day basis.

6.4.3 Phase-III: Larval Rearing on commercial pellet diet

With completion of Phase-II larval rearing, the weaning of larvae on grounded commercial feed would be complete. The larvae would attain an average size of around 20-25 mm TL and weight 2.0-2.5 g. Now, larvae are to be weaned on commercial pellet diet so that fouling of water and cost of feeding is minimized. Now, the larval density has to be reduced to 1.5 larvaeL^{-1}. It is better to either change the larval tank or the same tank may be cleaned thoroughly after taking out all larvae to avoid accumulation of pathogens on the wall of the tank (Fig. 37). To begin feeding with pellet diet, size of pellet given to be 0.4 mm that can be increased to 0.8 mm and 1.2mm respectively with increase in the size of larvae. It is better to mix little part of small size pellet diet with advancement of pellet size during feeding so that small size larvae can also feed easily. The larvae are trained to feed pellet size of 1.2 mm within 10 days (Fig. 38). Formulated diet @ 5-6% bw support higher growth, survival and better control over cannibalism. Once the larvae are totally weaned on pellet diet, the rate may be gradually reduced to 4-5%. The larvae provided feed 2-3 times in a day give better performance in terms of weight & length gain, FCR, SGR and PER. So weaned fingerling will now ready for stocking in either fingerlings rearing tank/ pond or grow-out pond, where they could exclusively reared on pellet diet with negligible cannibalism. Chart showing cost-effective larval feeding plan and regime for different age larvae is given in Figure 39.

Fig. 37: Larval rearing tanks of 300 litre

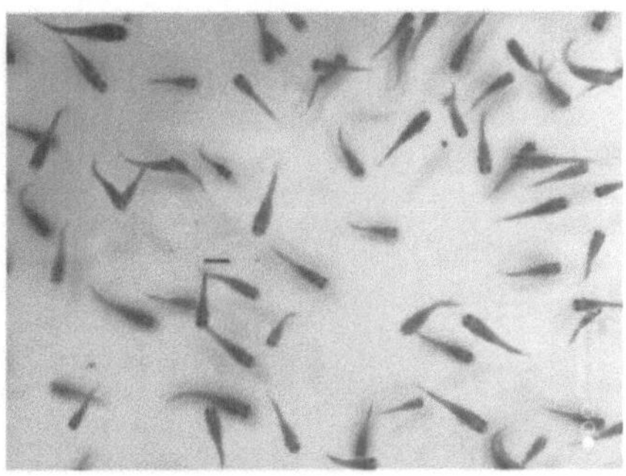

Fig. 38: Fry attained deep red colouration

<table>
<tr><td>III Feeding</td><td>BSN</td><td>BSN</td><td>Moina or fine plankton</td><td>Mixed Plankton</td><td rowspan="3">Fish mash+grounded formulated feed</td><td rowspan="3">Pellet diet</td></tr>
<tr><td>II Feeding</td><td>BSN</td><td>Moina or fine plankton</td><td>Mixed Plankton</td><td>Fish Mash</td></tr>
<tr><td>I Feeding</td><td>BSN</td><td>Moina or fine plankton</td><td>Mixed Plankton</td><td>Fish Mash</td></tr>
<tr><td></td><td>1-7 days</td><td>8-15 days</td><td>16-25 days</td><td>26-28 days</td><td>29-35 days</td><td>36-45 days</td></tr>
</table>

Fig. 39: Chart showing cost effective larval feeding plan and regime for different age larvae

7

Fingerling Production

Stocking fingerlings in grow-out ponds increases survival rates, cut-down culture duration, better feed management and reduce risk-factors on account of natural as well disease occurrence. In pond culture, good amount of fry-size seed suffer predation from enemies such as snakes, birds and predator fishes, if any and mortalities due to disease occurrence as smaller size fish is more vulnerable. However, fingerling stocking reduces predation loss to a greater extent, increase survivals, allow better feed management and highe production. This is truer with carnivorous fishes and particularly those observe cannibalism. Qin and Fast (1996) stated that the rate of cannibalism is higher in small size (TL) predators than the larger in case of striped murrel, so stocking large size fingerling in culture ponds reduces chances of ICC and hence recommended for grow-out production.

7.1 Traditional Methods of Fingerling Rearing

Wee (1981) has defined traditional system of murrel farming in Thailand in which seed of murrel was procured from wild resources and raised to large fingerling size before stocking. Two type of practices being used to raise fingerlings. In the first one, the pond(s) designated for fingerling production are stocked with brooders of SFs so they could spawn in the stocking pond(s) few days before the stocking of striped murrel fry and could directly feed upon the juvenile SFs. In second system, the fingerlings were reared to accept small-size fish (SF) along with rice bran to cut down cost of feeding on production of fingerling. In recent, more systematic approaches have been made to wean the seed collected from the wild resources on inert diets totally. The protocol adopted for weaning the fingerling on inert diet includes initial feeding with cut pieces of SFs and then gradual supplementation of inert diet till fingerlings start relishing inert diet completely.

However, the major disadvantages of using SFs are unpredictable survival and the so produced fingerling continues to have cannibalism habit that continued as long as the culture period. Feeding fry or fingerlings with SFS brought from outside also creates number of problems such as poor quality of SFS due to poor storage, risk of disease introduction and outbreaks, environmental pollution and higher FCR (4.2) in comparison to pellet diet (1.68) (So *et al.*, 2012). With availability of hatchery produced seed, weaning seed on inert diet is being promoted in some of the hatcheries in recent (*Haiwen et al., 2014*). Hatchery produced seed is more comfortable to wean on inert diet in comparison to seed procured from wild resources, as in earlier case the seed is acclimatized to the hatchery conditions and fed on exogenous diets since hatching.

7.2 Fingerling Rearing Systems

Fingerling rearing of hatchery produced advanced fry can be undertaken in small earthen ponds, hapas (Fig. 40), cement tanks (Fig. 41), FRP (Fig. 42) (capacity 500-5000 L), cages and pens. Modern systems of farming such as Flow-through, RAS and Biofloc systems are highly suitable for fingerling rearing of this species.

Fig. 40: A battery of hapa

Fig. 41: Cement tanks

Fig. 42: FRP tanks of 500 L

7.3 Rearing Protocols

A water depth of 2.5-3.0 feet is ideal. Tanks may be provided free-board of about one foot to prevent fish escape due to jumping. The fry and fingerling stages ofs striped murrel are surface dwellers and easily becomes the victim of predatory birds. Therefore, the pond/tank needs to be covered from top with wide meshed (mesh size 2.0-2.5 inch) netting (HDPE gill netting) that prevent entry of predatory birds. Forty to forty-five days old larvae weaned on formulated diet to be stocked in these rearing system.

Fig. 43: 40 days old hatchery reared seed for stocking in fingerling rearingpond

The seed may be stocked @ 30-50 larvae per cubic meter under pond and hapa system, 250-300 larvae per cubic meter in biofloc system, whereas stocking density in flow-through and RAS systems can be kept higher up to 500-1000 per cubic meter. They are fed initially on floating pellet diet (pellet size 1.2 mm; CP 40%, lipid 8%) at 5-6 % bw for about 30 days and later on with pellet diet of size 1.8 mm at 4-5% bw till they attain size of about 3-4 inches. The water quality to be adjusted near to normal parameters as detailed in section 6.2. Rearing in hapa system requires change of hapa on weekly basis to off shoot chances of pathogen development. The tank system has to be cleaned of all debris either on daily or alternate day basis with replacement of 30-40% tank water. Shooters, if any have to be removed from the stock at the earliest. The fingerling will be ready for stocking in the grow-out system within 45 to 60 days rearing with over 80-90 percent survival (Fig. 43). The fingerling so produced will be completely weaned on artificial diet and can be cultured on formulated diet during grow-out phase with survivals of over 90%.

Fig. 44: Advanced fingerling ready for stocking in grow-out ponds

7.4 Experimental Rearing in Earthen Ponds

Rahman *et al.* (2013) undertook fingerling rearing in 9 earthen ponds of size 0.012ha (depth 0.8m) for 8 weeks in Bangladesh. The early fry (size 1.17$\pm$0.18cm, 0.15$\pm$0.03g) collected from natural resources were fed BSN for 10 days and stocked in the nurseries in three densities of 1.5, 2.0 and 2.5 lakh per hectare. The fish were fed a mixture of fish meal+mustard oil cake in 1:1 @ 5-8 percent bw together with additional feed of chopped trash fish @ 5percent bw on daily basis. The performance of fish was found best in terms of growth in length, weight, SGR, FCR and survival with stocking density of 1.5 lakh fry per hectare.

7.5 Experimental Rearing in Cement Tanks

Experimental rearing of striped murrel fry to fingerling size was undertaken in two identical cement tanks (size 5x3x1.2m) at Lucknow (U. P.), India (authors unpublished data). The climate of the area is warm sub-tropical with extremities of hot summer (March-November) and cold winter (December to February). The cement tanks were provided soil base of 6 inch depth, so as to provide natural climate to the fish. The tanks were filled with tube well water to a depth of 0.5m. The average water temperature during the rearing period was 27$\pm$1.5°C. The water quality of the tank water was pH 8.1$\pm$0.3, total alkalinity 220$\pm$10 mgL^{-1}, DO 5.5$\pm$1.6 mgL^{-1}, free Co_2 8.0$\pm$3.1mg L^{-1}. Each tank was stocked with 225 hatchery produced striped murrel fry (weight 2.2$\pm$1.05g, length 6.7$\pm$1.0cm,) that equals to stocking density of 30 frym^{-3} (equivalent to 3.0 lakh fry per hectare metre). The fish were fed same floating pellet diet (CP 45%, lipid 10%, fibre 3.5; pellet size 1.2mm) which was used for rearing the fry in the hatchery, two times a day. The fingerling attained weight of 23.74$\pm$5.5g and length of 14.1$\pm$9.7mm in 30 days, which is a good size for stocking fingerling weaned on pellet diet in the grow-out ponds. The details of length-weight and growth parameters are given in Table 6. The survival was observed 98 percent at the end of the experiment, which was remarkable. The rate of SGR was very high (3.522 percent), which may be due to availability of larger rearing space to the larvae immediately after stocking fry in the cement tank and higher water temperature in outdoor tanks. Coefficient of determination (R-square, 0.911) indicated allometric growth but exponent of 'b' (3.522) and Ponderal Index (K, 0.844) indicated robust health of the fingerling in comparison to larger sizes in this species. The data indicated that fingerling of striped murrel can be reared on floating pellet diet in outdoor cement tanks with high survival rates and good growth within 30 days time at average temperature of 28$\pm$1.5°C.

Table 6: Growth parameters of striped murrel fingerling reared from hatchery produced fry in outdoor cement tanks.

Parameter	October 01, 2019 (stocking size)	October 31, 2019 (Harvest size)
Water temp.(°C)	-	27±1.5
Total length (cm)	6.70±1.0	14.1±9.7
Max. length (cm)	9.2	16.5
Min. length (cm)	5.3	11.9
Mean weight (g)	2.2±1.05	23.74±5.47
Max. weight (g)	5.6	38.0
Min. weight (g)	1.1	13.0
'a'	1.998	-2.330
'b' (exponent)	2.805	3.216
SGR (%) per day	-	3.522
R^2 (Coefficient of determination)	0.950	0.911
K (Ponderal Index)	0.736	0.844
Survival (%)	-	98

7.6 Experimental Rearing in Biofloc Media

Fingerling rearing of striped murrel on experimental basis was undertaken in indoor biofloc system at Lucknow, India (author"s unpublished data). The climate of the area was warm sub-tropical as detailed in the previous section. Biofloc media was developed in two identical round tanks of dia 2.5m and height 0.9m (depth of media 0.7m and volume 3400 litre). Each tank was stocked with 700 hatchery produced fry (weight 1.94±0.71g and length 6.29±0.81cm) of striped murrel weaned on floating pellet diet on November 1, 2019 (with setting of winter season) so as to get stocking density of 200 frym^{-3}. The target for rearing was to get fingerling of about 140-150mm length during winter season, which is ideal for stocking in grow-out ponds for fingerling weaned on pellet diet during spring. The fish were fed floating pellet diet (CP 45%, lipid 10% and fibre 3.5%) @ 1-4%bw 1-2 times a day, depending on feed acceptability during cold climate. It was observed that the fingerling attained weight of 23.74±5.4g and length of 14.12±0.97cm (pooled data of 50 fishes) with 98 percent survival in 5 months time (Table 7). The reason of slow growth was assessed to be the low water temperature as SGR, R2 and „b" values were initially good in the month of November but declined considerably during winter months. However, growth parameters improved well in March when temperature of water also improved from lowest of 15±1.0 to 21±2.0°C. The most interesting part was high survival of 98 percent at the end of experiment. The experiment thus revealed that fingerling rearing of striped murrel can be undertaken in biofloc system (indoor conditions) with high rate of survival during winter season. Though growth was slow in winter season, but it has opened possibility of raising fingerling of this species during winter climate using pellet diet. The availability of fingerling during March or maximum during April months can provide opportunity of initiating grow-out production with setting of summer climate to get maximum growth during complete summer season.

Fig. 45: Fry to fingerling rearing in biofloc media

Table 7: Growth parameters of striped murrel fingerling production during winter climate in indoor biofloc system

Parameter	November 1, 2019 (stocking size)	Nov. 30, 2019	Dec. 31, 2019	Jan. 31, 2020	Feb. 29, 2020	Mar. 31, 2020
Water temperature (°C)	27±2.0	25±2.0	17±1.5	15±1.0	18±2.0	21±2.0
Total length (cm)	6.29 ±0.81	9.42 ±0.77	10.80 ±0.93	11.30 ±0.78	12.00 ±1.12	14.06 ±1.02
Max length (cm)	8.2	10.7	13.2	13.1	14.3	16.5
Min length (cm)	4.9	8.0	8.5	9.5	9.5	12.5
Mean weight (g)	1.9 ±0.7	7.48 ±1.75	10.53 ±2.40	12.5 6±2.54	14.93 ±3.53	23.2 6±5.83
Max weight (g)	3.9	12.0	18.0	19.0	2.0	38.0
Min weight (g)	1.1	5.0	16.0	7.0	8.0	16.0
SGR % per day	-	1.967	0.481	0.250	-0.086	0.730
R^2 (Coefficient of determination)	0.936	0.858	0.768	0.899	0.826	0.929
K (Ponderal Index)	0.792	0.895	0.836	0.868	0.611	0.835
Proportionality constant 'a'	-1.679	-1.680	-1.268	-.1949	-1.605	-2.318
Exponent 'b'	2.454	2.614	2.210	2.887	2.567	3.201
Survival (%)						98

7.7 Experimental Rearing in Recirculatory Aquaculture System (RAS)

Saputra *et al.* (2018) have estimated the suitable stocking density of striped murrel juveniles (juvenile size 3.41±0.39cm, 0.18±0.07g) in RAS with respect to growth and survival and observed insignificant differences in growth and survival with maximum stocking density of 3 juvenileL^{-1} and survivals of 92.5-94.6 percent in 42 days rearing period.

8

Grow-Out Production

The farming of striped murrel is quite old in many south-eastern and south Asian countries including India under traditional farming system. Wee (1981) has given an account of traditional farming of striped murrel in Thailand, where seed of wild stock was stocked in very high density (75-460 numbers/m^2) in ponds. Freshwater small fish (FSF) were fed to the seed during pre-monsoon and monsoon seasons and subsequently marine small fish (MSF) mixed with rice bran to satiation level. The fish grew to 700-1000 g in 7-10 months with an FCR of 6.5:1 and production of 9.0-15.6 kg/m^2. In another culture trial, *Diana et al.* (1985) have described a poor survival of 9-13%, when they stocked seed @ 40-80 number/m^2 collected from wild sources and fed them on FSF for 9-11 months. There are records of backyard production of this species also (Wan *et al.*, 1992). The major problems with traditional farming were poor & uncertain survivals, problems of getting SFs and occurrence of diseases due to poor management protocols.

Therefore for a scientific farming of this species, the major bottlenecks are the non- availability of hatchery seed, species & size specific feed, cannibalism at all stages and disease occurrence. These bottlenecks have been addressed to a greater extent by the researchers in recent and protocols for hatchery seed production and cannibalism reduction have been simplified to a greater extent with the availability of commercial pellet diets to boost its farming. Vietnam has been pioneer in the production of farmed striped murrel in recent, following new initiatives and based on its success, neighbouring countries such as Indonesia, Camobodia, Thailand, Myanmar, etc have also taken initiatives to boost production of this high value species in their countries.

Striped murrel is also a valuable fish species in India particularly in the states of Andhra Pradesh, Telangana, U.P, Bihar, Odisha, West Bengal, etc and farmed under traditional methods of farming. With burgeoning demand in parts of Telangana and Andhra Pradesh, few farmers have taken up farming of this species on the lines of new initiative taken up in other countries. The experimental culture under new farming systems such as flow-through, biofloc and RAS could further help in intensive farming practices while minimizing the chances of disease occurrence.

Based on the available research and author"s own experience, the grow-out production of this species carried out in recent under different farming systems is herewith summarized and a protocol has been suggested for the farming of striped murrel.

8.1 Farming in Earthen Ponds

So far, there are no standardized protocols for grow-out production in respect of striped murrel. The overall production depends on the adopted criteria such as (1) source of seed, whether from wild resources or hatchery bred, (2) whether seed has been weaned on a particular diet or not; if yes, whether on meat (fish) or formulated diet prior to stocking, (3) whether size-grading of seed has been done, (4) the type of feed to be given during grow-out period, and 5) stocking size of seed and density. Case studies of grow-out production undertaken by different workers are given as under.

Case study-I: So *et al.* (2012) carried out a comparative study to assess the growth performance of striped murrel fed with FSF and formulated pellet feed under hapa (size 5.0X3.0X1.5m) condition fitted in a pond of $2000m^2$ in Cambodia. The hapa were stocked with 60 days old striped murrel fingerling @ 50 fingerlings per m^{-2} and reared for 10 months. Fishes fed with FSF showed higher growth (average 468 g) in comparison to pellet fed (CP 40%) fishes (average 315 g) with no significant difference in survival (60 and 56%).

Case study-II: Rahman *et al.* (2012) carried out grow-out of striped murrel in earthen ponds (size 0.048ha) with 3 stocking densities 5000, 6250 & 7000 fingerlings (weight 17.63 ± 1.23g and length 13.21 ± 0.52cm) per hectare collected from wild sources and fed them with fish meal (30%) & rice bran (70%) @ 3-6% bw and also with fish @ 2-3%bw on every alternate day for 8 months. The best production of 1479.4 kg/ha with 67% survival was obtained with stocking density 5000/ha.

Case study-III: Haiwen *et al.* (2014) have reported large-scale commercial farming of striped murrel in Myanmar during 2012 and 2013. A total of 18 earthen ponds (total area 14.4 ha) were stocked with 5.4 lakh fingerling of average weight 6.2g (stocking density 37, 500/ha) procured from wild resources in the year 2012. The fingerling weaned on floating pellet diet (CP 39-43%), were fed the same diet during entire culture period. In a culture of 342 days, production of 48.03 ton fish was obtained by them with 85.3% survival and FCR of 1.3. The weight range of fish was between 500-1000 gm. In subsequent trial during 2013, the culture was increased to 40 ponds (total area 36 ha) following similar stocking density and feed protocol. A production of 107 ton in 182 days was obtained by them with FCR of 1.11 with fish weighting in the range of 500-800g.

Case study-IV: Farhana *et al.* (2016) cultured striped murrel in 3 earthen ponds each of 30 decimal (1200m^2). The ponds were stocked with fingerling of striped murrel (length 11.75cm and weight 45g) collected from wild resources @ 10,000/ha. Feeding was done with introduction of live fry of bata (*Labeo bata*), mrigal (*Cirrhinus mrigala*) and silver carp (*Hypophthalmicthys molitrix*) @ 1-3% bw of striped murrel and also fed mixture of rice bran, mustard oil cake and fish meat (10:10:1 ratio) @ 3-5% bw to serve as feed for all the species. The fishes were harvested after 180 days with a gross production of 2143.7 kg with average size of 850g and 97% survival. Besides, above production, 500kg additional production of other species was also obtained by them.

Case study-V: Rao (2020), a successful murrel farmer of Machlipatnam, Andhra Pradesh, India; in a recent study has demonstrated farming of striped murrel in an earthen pond of size 0.6 ha with stocking density of 26,000 juveniles (40g) fed on pellet diet. In a culture of 8-months, a production of 20.93 ton (average size 850g) with 95% survival and FCR of 1.5 was obtained by him.

Protocol suggested for farming in earthen ponds

With the outcome obtained from the case studies given above, the authors suggest protocol for grow-out production under pond conditions as under.

Small ponds (<0.4 ha) are better for murrel farming from management point of view (Fig. 44). A water depth of 1.0-1.5m is ideal to minimize temperature fluctuations and energy stress during obligatory aerial respiration by the fish. Both new and old ponds (after drying) may be used, however the pond should have sufficient gradient to draw water at one end during harvesting with a free-board of about 45 cm for preventing fish escape due to jumping and monsoon flashes. It is better to tie a screen all round the pond embankment during rainy season for preventing fish escape. The pond(s) is to be fenced from top with netting (HDPE gill netting of mesh bar 2.0-2.5 inch) to avoid fish predation by the birds at least for a period till fish attains size of about 9-10 inches.

Fig. 46: Earthen pond for grow-out production

This protocol is suggested with 90 days old fingerling of striped murrel weaned on pellet diet. The pond has to be manured with cowdung (approximately 500 kg/ha) and fertilizer (Diammonium phosphate approximately @ 30 kg/ha) to maintain turbidity of around 30-40 cm (Secchi-disc transparency). Five to ten days after manuring, the pond has to be stocked with 90 days old fingerling, previously weaned on artificial pellet diet, @ 50,000-60,000 numbers per hectare. The fingerlings are size graded to the extent that smaller fish should not be less than half the size of larger ones. The fingerling are to be fed floating pellet feed (CP 40%, lipid 6-8% and ash 3.5%) 2 time per day @ 3-2% body weight (reduce with increase in size). Standard pond water quality has to be maintained to avoid chances of disease occurrence. The fish may attain an average weight of 750 gm with an average survival of about 90%. A production of 33-40 ton per hectare in 8-9 months with around 90% survival may be expected with FCR of 1.5.

8.2 Cement Tanks

The grow-out production of giant murrel (*Channa marulius*) has been attempted by Yadav *et al.* (2014) in cement tank and had raised possibility of farming of murrels under such system. However, there is no available data on grow-out production of striped murrel in cement tanks. In order to observe the grow-out potential of this species in cement tanks, the authors undertook experimental culture in two identical cement tanks of size 15 m^2 (5.0m X 3m) provided with 6 inch soil base during 2019-20 (Fig. 45). The tanks were filled with bore-well water up to a depth of 0.7m and covered with netting from the top to avoid bird predation. Each tanks was stocked with 210 hatchery produced striped murrel fingerling (length 14.12$\pm$0.97 cm and weight 23.74$\pm$5.42 g), which were previously weaned on pellet diet (CP 45%, lipid 10%, fibre 3.5%) at the same farm facility thus making a stocking density of 20 fingerling.m^{-3}. The fingerlings were stocked on 1st November, 2019 with start of winter season and cultured till 31st August 2020. Thus ten months culture includes 4 months winter season (November-February) and 6 months summer (March-August). The fish were exclusively fed on floating pellet diet (CP 40%, lipid 8% and fibre 3.5%) ad- libitum two times in a day (morning & evening). However, the feeding suffered during April and May months due to lock-down of the country in view of Covid19

Fig. 47: Cement tanks used for grow- out production

Fig. 48: 5-month old striped murrel grown in cement tank

Pandemic in India and could be provided only once in a day. The tank water was partly changed every month except during lockdown period. The data on length- weight and growth were taken month-wise and survivals in March and August by completely draining the tank. The LWR, growth and survival data were evaluated for 10 months rearing period November, 2019 to August, 2020. The details are presented in Fig. 49 and Table 8.

The pooled LWR data of both the tanks revealed a linear and uniform growth trend with emergence of 2-3 numbers of outliers in each tank. The survival was found 85 percent and the mortality was observed only during severe winter period. Outliers were not observed to cannabalise their siblings as no mortality was observed after March. The fish attained a weight of 203±59g and 28.08±2.81cm from an initial weight and length of 23.74±5-.42g and 14.12±0.97cm respectively. The growth was poor in winter in comparison to summer season and improper feeding during Covid19 period was reflected in the overall growth of fish at the end of culture period. However, an average weight of 203±59g in 10 months period empassing 4 months winter with high stocking density of 20m³ in cement tank was remarkable, considering the commerciable importance of this species.

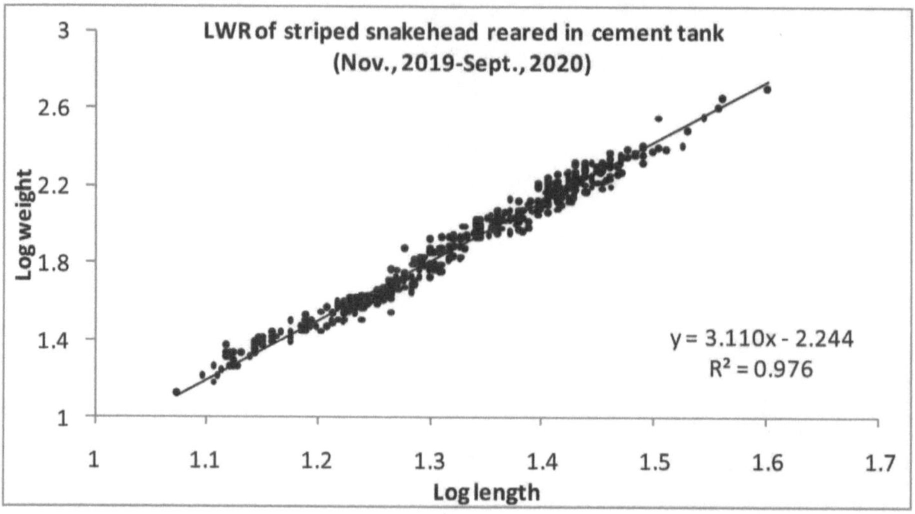

Fig. 49: Trend of LWR of striped murrel reared in cement tank for 10 months

8.3 Biofloc Technology (BFT)

The grow-out production of striped murrel in BFT was undertaken in two rectangular cement tanks of 10 tonne capacity by the authors during April-August 2020 (authors unpublished data) (Fig. 50). The biofloc media was developed in the tanks and 650 fingerling (65 fingerlingm³, size 23.26±5.83g, 14.06±1.02cm), reared from hatchery produced fingerling size in biofloc media separately, were stocked in each tank in April, 2020. The fish were fed on floating pellet diet (CP 45%, lipid 10%, crude fibre 3.5%; pellet size 1.2mm) @ 2-3 percent body weight. The fishes were sampled (50 number) every month for growth data and survival in August.

Fig. 50: A view of biofloc grow-out tanks

Five months culture period revealed survival of 98 percent and the LWR indicated a uniform growth without development of any outlier in the population in both tanks.

Table 8: LWR and growth indices of striped murrel reared in cement tank

Parameter	1 Nov., 2019 (Stocking size)	1 Dec., 2019 (30 days)	1 Jan., 2020 (60 days)	1 Feb., 2020 (90 days)	1 Mar., 2020 (120 days)	1 April 2020 (150 days)	1 May 2020 (180 days)	1 June 2019 (210 days)	1 July 2020 (240 days)	1 August 2020 (270 days)	1 Sept., 2020 (300 days)
Water temperature °C	22±2.0	15.0±1.5	16.0±2.0	19.0±2.5	24.0±2.0	26.0±2.0	27±1.5	29.0±2.5	29.0±2.0	28.5±2.0	27.5±1.5
Mean total length (cm)	14.12 ±0.97	17.20 ±1.55	18.05 ±1.55	18.02 ±0.95	18.76 ±1.40	20.58 ±0.98	22.90 ±1.73	24.32 ±2.25	26.01 ±2.29	27.76 ±2.23	28.08 ±2.81
Max length (cm)	16.5	21.0	21.3	21.3	21.5	22.6	27.0	35.0	36.2	36.5	40.0
Min length (cm)	11.9	14.6	16.5	16.5	16.4	19.3	19.0	22.0	22.5	23.5	25.0
Mean weight (g)	23.74 ±5.42	38.25 ±10.67	44.42 ±7.08	43.46 ±7.34	47.22 ±1.40	74.74 ±7.07	106.18 ±18.00	118.44 ±2.23	138.24 ±48.58	180.12 ±48.97	202.76 ±59.14
Max weight (g)	38.0	67.0	68	68.0	74	94.0	164.0	358.0	402	446	500.0
Min weight (g)	13.0	24.0	36.0	35.0	30	62.0	74.0	84.0	86	102	152.0
Proportionality constant 'a'	-2.330	-1.916	-1.590	-1.664	-2.171	-1.841	-0.794	-1.725	-2.241	-1.470	-0.953
Exponent 'b'	3.216	2.823	2.574	2.626	3.013	2.826	2.072	2.733	3.087	2.475	2.245
Coefficient of determination 'R^2'	0.911	0.921	0.876	0.776	0.935	0.943	0.887	0.912	0.883	0.816	0.854
Ponderal Index 'K'	0.844	0.750	0.754	0.742	0.715	0.857	0.884	0.822	0.785	0.842	0.915
SGR (%) per day	-	0.655	0.242	0.032	0.102	0.670	2.072	0.123	0.213	0.391	0.167
Survival (%)	-	-	-	-	-	-	-	-	-	-	85

The fish attained average weight of 149.60±28.58g and length 25.44±1.81cm during this period, which is remarkable and revealed culture potential of this high value species in biofloc system. The growth rate was found better in the months of June and July. With this growth rate, it is expected that the fish fish may attain an average weight of about 200-225g in 12 months culture period at this stocking density. The trials in China with *C. argus* has also demonstrated suitability of biofloc media for raising 300g size fish in 10-20 tonnes tanks.

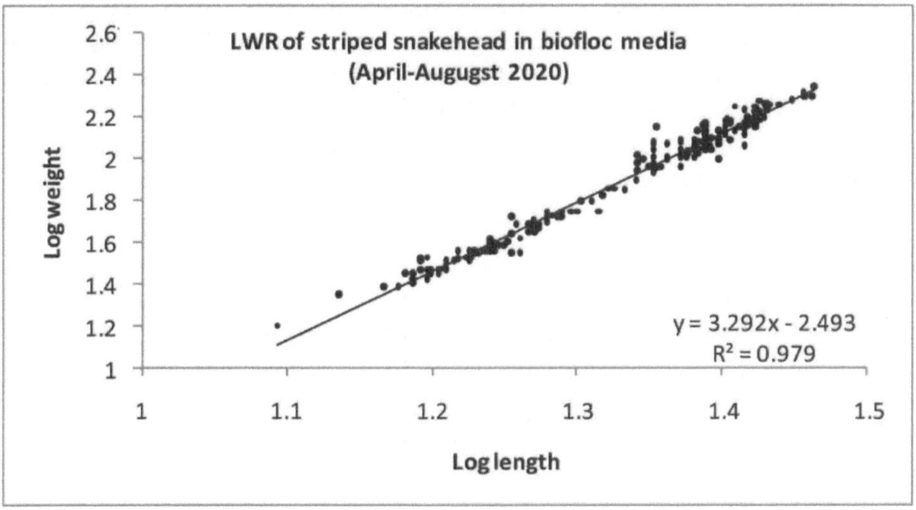

Fig. 51: LWR of striped murrel in biofloc grow-out system (5 months data)

Table 9: LWR and growth of striped murrel reared in biofloc treatment (BFT)

Parameter	April 1, 2020 Stocking size	April 30, 2020 (30 days)	May 31, 2020 (60 days)	June.30, 2020 (90 days)	July 31, 2020 (120 days)	August 31, 2020 (150 days)
Water temperature (°C)	21±2.0	22±2.0	24±2.0	26±1.5	27±1.5	28±1.5
Mean Total Length (cm)	14.06 ±1.02	16.50 ±0.80	18.28 ±1.75	23.96 ±1.36	25.31 ±1.09	25.44 ±1.81
Max length (cm)	16.5	18.2	21.5	26.8	29.0	29.1
Min length (cm)	12.5	15.0	12.4	20.4	23.0	22.0
Mean Weight (g)	23.26 ±5.83	32.60 ±3.93	45.82 ±11.57	108.26 ±19.00	133.88 ±21.63	149.60 ±28.58
Max weight (g)	38.0	43.0	72.0	156.0	196.0	216.0
Min weight (g)	16.0	24.0	16.0	62.0	100.0	94.0
SGR% per day	-	0.502	0.449	1.234	0.311	0.147
R^2 (Coefficient of determination)	0.929	0.785	0.946	0.850	0.806	0.851

K (Ponderal Index)	0.835	0.943	0.749	0.786	0.825	0.908
'a' (Proportionality constant)	-2.318	-1.208	-0.1.814	-1.926	-2.278	-1.312
'b' value	3.20	2.23	2.74	2.86	3.27	2.47
Survival (%)	-	-	-	-	-	90

8.4 Cage Culture

Cage culture for farming of striped murrel has been practiced only at experimental stage. Experimental culture of striped murrel was undertaken in 3 bamboo cages of size 18 m^3 in Laos PR by Living Aquatic Resources Research Centre (LARReC) of NACA (*LARRec, 2002)*. The cages were stocked with 10 cm size juvenile fishes collected from wild resources and stocked @ 56 juveniles/ m^3 for 7 months. The fish grew to an average size of 500g with 97% survival and FCR of 3.5 when fed on small dry fish.

Muthumainnah (2013) has also described the performance of striped murrel in bamboo cages in Cambodia. Six cages of size 2 x 1.5 x 1.5 m, were stocked with fingerling collected from wild resources having length 6.2cm & weight 2.18g at three stocking densities 50, 100 and 150 numbers per m^3 and fed with pellet diet (CP 38%). The best performance was obtained with stocking of 50/m^2 with a growth rate of 16.32 cm & 48.25 g with production of 2412.5 g/m^3 in 12 weeks.

8.5 Mixed Farming of Murrel and Carps

Intensive farming of the murrel along with the carps is preferred in China as carps use to convert the effluents produced during murrel farming. Qinfen *et al.* (2013) described the outcome of mixed farming using hybrid murrel (female *C. asiatica* ♀ X *C. argus* ♂), silver carp (*Hypophthalmicthys molitrix*) and crucian carp (*Carassius carassius*) undertaken in a pond of 0.27 ha (depth 2.5 m). The pond was stocked with seed of hybrid murrel (2,00,000 nos. of size 66g), silver carp (2800 nos. of 33g) and crucian carp (7300 nos. of 10.5g) and fed pellet diet @ 3-5% split into 3 feedings. The fish attained weights of hybrid murrel 700g, silver carp 1150g and crucian carp 210g with respective survivals and yield of 70% (26,732 kg), 87% (650 kg) and 46% (195 kg) in 190 days.

8.6 Nutritional Requirement for Grow-out Production

The crude protein requirement of striped murrel at grow-out phase (size 31g to about 200g) has been estimated by the saturation kinetic model to be 51.6%

DM for growth and 52.1% DM for protein deposition (Hua *et al.*, 2019). According to So *et al.* (2012), the crude protein demand for striped murrel is around 50 percent. Increasing dietary lipid resulted in a significant increase of body lipid contents, indicating dietary lipid was directed towards body lipid deposition rather than towards protein-sparing. However, dietary level of lipids >19% is detrimental on growth on this species (Hua *et al.*, 2019). They further elaborated that higher levels of lipid has no sparing property in terms of dietary protein.

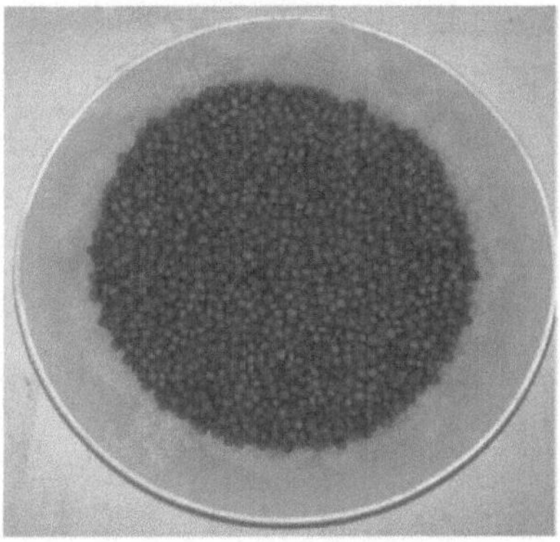

Fig. 52: Floating Pellet diet for grow-out

References

Abol-Munafi A.B., B.M. Tam, M.A. Ambak and P. Ismail (2004). Effect of different diets on growth and survival rates of murrel (*Channa striata* Bloch, 1797) larvae. *Korean J. Biol. Sci*, 8: 313-317.

Ali A.B. (1999). Aspects of the reproductive biology of female murrel (*Channa striata*, Bloch) obtained from irrigated rice agro ecosystem, Malaysia. *Hydrobiol*, 411: 71-77.

Alikunhi K. H. (1953). Note on bionomics, breeding and growth of the murrel *Ophiocephalus striatus* (Bloch). *PNAS, India.*, 38: 10-20. .

Aliyu-Paiko M., R. Hashim and A.C. Shu-Chien (2010). Influence of dietary lipid/protein ratio on survival, growth, body indices and digestive lipase activity in murrel (*Channa striatus*, Bloch 1793) fry reared in re-circulating water system. *Aquacult. Nutr.*, 16: 466-474..

Ayyappan, S., Sudhir Raizada and A.K. Reddy (2001). Captive breeding and culture of new species of aquaculture. NBFGR-NATP Publication No. 3, Key Paper no. 1. Publisher: National Bureau of Fish Genetic Resources, Lucknow, India.

Babiak, I., S.N.M. Mandiki, K. Ratsinjomanana and P. Kestemont (2004). Initial weight and its variation in post-larval Eurasian perch affect quantitative characteristics of juvenile cohorts under controlled conditions. *Aquaculture*, 243: 263–276.

Baras E. (1998). Bases biologiques du cannibalisme chez les poisons. *Gh Ethol.*, 18:53-98.

Baras E. M. Nsao, M.Y.J. Maxi, D. Jeandrain, J.P. Thome, P. Vandewalle and C. Melard (2000). Sibling cannibalism in dorada under experimental conditions. I. Ontogeny, dynamics, bioenergetics of cannibalism and prey size selectivity. *J Fish Biol.*, 57: 1001-1020.

Baras, E. and M. Jobling (2002). Dynamics of intracohort cannibalism in cultured fish. *Aquacult. Res.*, 33: 461-479.

Baras, E., P. Kestemont and C. Mélard (2003). Effect of stocking density on the dynamics of cannibalism in sibling larvae of *Perca fluviatilis* under controlled conditions. *Aquaculture*, 219: 241–255.

Baras, E., J. slembrouch, C. Cochet, D. Caruso and K. Lengendre (2010). Morphological factors behind the early mortality of cultured larvae of the Asian catfish, *Pangasianodon hypophthalmus*. *Aquaculture*, 298: 211-219.

Baras É., D.V. Silva Del Aguila, G.V. Montalvan Naranjos, Remi Dugue, Fred Chukoo, Fabrice Duponchelle, Jean-Francois Renno, Carmen Garcia-Darilla and Jesus Nunez (2011). How many meals a day to minimize cannibalism when rearing larvae of the Amazonian catfish *Pseudoplatystoma punctifer*? The cannibals point of view. *Aquat. Living Resour.*, 24: 379–390.

Baras E., R. Hafsaridewi, J. Slembrouck, Agus Priyadi and Yann Morean (2013). Do cannibalistic fish possess an intrinsic higher growth capacity than others? A case study in the Asian redtail catfish *Hemibagrus nemurus* (Valenciennes, 1840). *Aquac. Res.*, 45: 68–79.

Bhattacharya R. (1946). On the life history of *Ophiocephalus striatus* (Bloch). *Proc. Indian Sci. Congr.*, 33 (3): 125-29.

Bhatti M.F. (2010). Some studies on reproductive biology of sol (*Channa marulius*) in the Punjab, Pakistan. Ph.D. Thesis, Department of Zoology, Faculty of Sciences, Pir Mehr Ali Shah Arid Agriculture University, Rawalpindi, Pakistan.

Bilal A. P., M.A. Haniffa, D. Manikandaraja (2012). Induced ovulation and spawning of a stripped murrel murrel, *Channa striatus* (Bloch) under captive conditions. *J. Res. Ani. Sci.*, 1 (1): 33-39.

Bilal A. P., M.A. Haniffa, D. Manikandaraja and M. J. Milton (2013). Breeding behaviour and parental care of the induced bred striped murrel *Channa striatus* under captive condition. *Turkish J. Fisher. Aquat. Sci.* 13: 707-711.

Bilal A.P. and M.A. Haniffa (2016). Effect of feed quality on growth and survival of striped murrel, *Channa striatus* (Bloch, 1793) hatchlings. *Indian J. Geomarine Sci.*, 45 (1): 105-110.

Bobe, J.Y. and C. Labbe (2010). Egg and sperm quality in fish. *Gen. Comparat. Endocrinol.*, 165 (3): 535-548.

Breder, C.M. and D.E. Rosen (1966). Modes of reproduction in fishes. T.F.H. Publications, Neptune City, New Jersey, 941p.

Britz P.J. and A.G. Pienaar (1992). Laboratory experiments on the effect of light and cover on the behaviour and growth of African catfish, *Clarias gariepinus* (Pisces: Clariidae). *J. Zool.*, 227: 43-62.

Bui Minh Tam, Nguyen Thanh Phuong and Duong Nhut Long (2008). Effects of HCG dosages and injection methods on semi-artificial propagation of giant murrel (*Channa micropeltes*). Aquaculture and Fisheries College, Can Tho University, Vietnam (Tap chi Khao hoc (2): 76:81.

Chacko P.I. and G.K. Kuriyan (1947). On the culture of *Ophiocephalus marulius* (Hamilton), in the Coimbatore and Salem Districts, Madras. *Proc. Indian Sci. Congr.*, 37 (3): 180.

Changa, Yu-Han, Yu-Hung Linb, Chih-hao Hsiehc, Yi-Min Chena (2019). *Aurantiochytrium* dietary supplements reduce intra-cohort cannibalism among orange-spotted groupers (*Epinephelus coioides*) by modulating brain 5-HT and serum cortisol. *Aquaculture*, 502: 202-2011.

Chaudhry,S., S. De Alwis Goonatilake, S. M. Fernado and O. Kotagama (2019). *Channa striata. The IUCN Red List of Threatened Species* 2019: e.T166563A60591113.https://dx.doi.org/10.2305/IUCN.UK.20193.RLTS.T166563A605911 13.en.

Courtenay, W. R. Jr., J. D. Williams, R. Britz, M. N. Yamamoto and P. V. Loiselle (2004). Identity of introduced murrels (*Pisces, Channidae*) in Hawaii and Madagascar, with comments on ecological concerns. Bishop Museum Occasional Papers 77, Honolulu, 13 pp.

Das M., S.C. Chakraborty, F. Ahmed and R.K. Basak (1998). Predatory behaviour of a murrel fish (*Channa striatus* Bloch). *Bangladesh J. Fish. Res.*, 2 (2): 127-137.

Dayal R., P.P. Srivastava, A. Bhatnagar, S. Raizada, S. Chowdhary, A.K. Yadav and W.S. Lakra (2013). Captive spawning of the striped murrel, *Channa striatus* (Bloch) using sGnRH, in Gangetic Plains of India. *Proc. Natl. Acad. Sci., India, Sect. B Biol. Sci.* 83 (1): 65-70.

Dayan A. Perera, P. Phelps Ronald, Dang Diem Tuong, Miguel Zapata, Bui Minh Tam, William H. Daniels (2016). Use of Shrimp *Penaeus* spp. and Krill *Euphauisa* spp. in a feed training diet for striped murrel *Channa striata. Interntl. J. Fishe. Aquat. Stud.,* 4 (3): 12-15.

Devraj M. (1973). Biology of the large murrel, *Ophiocephalus marulius* (Ham.) in Bhavanisagar water. *Indian J. Fisheries*, 20 (2): 280-307.

Diana J.S., W.Y.B. Chang, D.R. Ottey and W. Chuapoehuk (1985). Production systems for commonly cultured freshwater fishes of Southeast Asia. Pages 75-79 in International Program Report, number 7, Great Lake and Marine Water Center, University of Michigan, Ann Arbor, Michigan, USA.

Dou Shuozeng, Tadahisa Seikai and Katsumi Tsukamoto (2000). Cannibalism in Japanese flounder juveniles, *Paralichthys olivaceus*, reared under controlled conditions. *Aquaculture*, 182 (1): 149-159.

FAO (2016). Species Fact Sheet *Channa striatus* (Bloch, 1793). Fisheries and Aquaculture Department, Food and Agriculture Organization of United Nations, Rome.

Farhana Tania, Md Ekramul Hasan, Abdullah-Al Mamun and Muhammad Shafiqul Islam (2016). Commercially culture potentiality of striped murrel fish *Channa striatus* (Bloch, 1793) in earthen ponds of Bangladesh. *Intel. J. Pure Appl. Zool.* 4 (2): 168-173.

Ferdausi Hosney Jannatul, Nirmal Chandra Roy, Mst Jannatul Ferdous, Mohammad Amzad Hossain, Md Mehedi Hasan, Trina Bristi Das, Mian Sohel, Iqbal Mohammed Mahbub, Munir Mohammad Bodrul and Mohammad Mosarof Hossain (2015). Reproductive Biology of Striped Murrel (*Channa striata*) from Natural Wetlands of Sylhet, Bangladesh. *Annals of Veterinary and Animal Science*, 2 (5): 162-169.

Folkvord A. (1991). Growth survival and cannibalism of cod juveniles (*Gadus morhua*): effects of feed type, starvation and fish size. *Aquaculture*, 97: 41-59.

Folkvord A. and H. Otterå (1993) Effects of initial size distribution, day length, and feeding frequency on growth, survival, and cannibalism in juvenile Atlantic cod (*Gadus morhua* L.). *Aquaculture*, 114: 243–260.

Fox L.R. (1975). Cannibalism in natural populations. *Annual Review of Ecology and Systematics*, 6: 87-106.

Francis, T., N. S. Ramanathan Athithan and H.F. Cheryl (2000). Induced breeding of murrel, *Channa striatus* using various inducing agents. *Fishing Chimes*, 19: 119- 121.

Ghaedi Alireza, Anamul Kabir Mohammad and Hashim Roshada (2013). Oocyte development and fecundity of murrel murrel, *Channa striatus* (Bloch 1793) in captivity. *Asian Fishe. Sci.*, 26: 39-51.

Ghaedi Alireza, Anamul Kabir Mohammad and Hashim Roshada (2016). Effect of lipid levels on the reproductive performance of murrel murrel, *Channa striatus*. *Aquacul. Res.*, 7: 983-991.

Haiwen Bai, Huang Shaoyu, U Than Lwin, U Tint Swe, Dong Qiufen, Zhang Song and Yang Yong (2014). The murrel fish: a success in Myanmar. AQUA Culture Asia Pacific Magazine (April-May): 20-23pp.

Haniffa M. A., J. S. Mohammed and T. Merlin Rose (1996). Induction of ovulation in *Channa striatus* (Bloch) by GnRH. *Fishing Chimes*, 16 (5): 23-24.

Haniffa M. A., K. Marimuthu and M. Muruganandam (2001). Mass breeding of the striped murrel *Channa striatus*. *Fishing Chimes*, 21 (10 & 11): 53-54.

Haniffa M. A., M. Nagrajan, K. Marimuthu and J. Arokiyaraj (2003). Embryonic and larval development of spotted murrel *Channa punctatus*. *Ind. J. Fishe.*, 50: 355-362.

Haniffa M. A., T. Merlin and J. S. Mohammed (2000). Induced spawning of striped murrel *Channa striatus* using pituitary extract, human chorionic gonadotropin and lutenizing hormone releasing hormone analogue and ovaprim. *Acta. Ich. Piscat.*, 30 (1): 53-60.

Hecht T. and A.G. Pienaar (1993). A review of cannibalism and its implication in fish larviculture. *J. World Aquacul. Soc.*, 24: 246-261.

Hecht T., S. Appelbaum (1988) Observations on intraspecific aggression and coeval sibling cannibalism by larval and juvenile *Clarias gariepinus* (Clariidae: Pisces) under controlled conditions. *J. Zool.*, 214: 21-44.

Hoelzer G.A. (1992). The ecology and evolution of partial-clutch cannibalism by parental Cortez damselfish. *Oikos*, 65: 113-120.

Hoff, F.H. and T.W. Snell (1989). Plankton Culture Manual. Florida Aqua Forms, Inc., Pisces Publishing Group, Devon, CT, 125 pp.

Hossain M.K., G.A. Latifa and M.M. Rahman (2008). Observation on induced breeding of murrel murrel, *Channa striatus* (Bloch 1793). *Intnl. J. Sustain. Crop. Prod.*, 3: 65-68.

Hseu, J.R. (2002). Effects of size difference and stocking density on cannibalism rate of juvenile grouper *Epinephelus coioides*. *Fish. Sci.*, 68: 1384-1386.

Hseu, J.R. (2003). Effect of exogenous tryptophan on cannibalism, survival and growth in juvenile grouper, *Epinephelus coioides*. *Aquaculture*, 218 (1-4): 251-263.

Hua Katheline, Wolfgang Koppe and Ramon Fontanillas (2019). Effects of dietary protein and lipid levels on growth, body composition and nutrient utilization of *Channa striata*. *Aquaculture*, 501: 368-373.

ICAR-CIFA (2016). Breeding and seed production of striped murrel. ICAR-CIFA Extension Series 26. Publisher: Director, ICAR-CIFA, Bhubaneswar.

Irmawati, J., Nadiarti Tresnati and L. Fachruddin (2019). Sex differentiation and gonadal development of striped murrel (*Channa striata* Bloch, 1793). IOP Conf. Series: Earth and Environment Science, 253: 1-8.

Ismail R.F., S.S. Assem, A. F. Fahmy, N. M. AbouShabana (2016). Reproductive biology, steroid and biochemical profiles of *Dentex dentex* ovaries in the Eastern Mediterranean in relation to histological structure Egypt. *J. Aquat. Res.*, 42: 149-160.

IUCN (2020) *The IUCN Red List of Threatened Species*: e.T166563A60591113. https://dx.doi.org/10.2305/IUCN.UK.2019-3.RLTS.T166563A60591113.en.

Izquierdo M.S., H. Fernandez-Palacios and A.G.J. Tacon (2001). Effect of broodstock nutrition on reproductive performance of fish. *Aquaculture*, 197: 25-42.

Katavić I., J. Jug-Dujaković and B. Glamuzina (1989). Cannibalism as a factor affecting the survival of intensively cultured sea bass (*Dicentrarchus labrax*) fingerlings. *Aquaculture*, 77: 135-143.

Kestemont P., S. Jourdan, M. Houbart, Charles Melard, Mihalis Paspatis, Pascal Fontaine, Armande Cuvier and Maria Kentouri (2003). Size heterogeneity, cannibalism and competition in cultured predatory fish larvae: biotic and abiotic influences. *Aquaculture*, 227: 333-356.

Khan H. (1924). Observations on the breeding habits of some freshwater fishes in Punjab. *J. Bombay Nat. Hist. Soc.*, 29: 958-962.

Kilambi, R. V. (1986). Age, growth and reproductive strategy of the murrel, *Ophiocephalus striatus* Bloch, from Sri Lanka. *J. Fish Biol.*, 29: 13-22.

Kokokiris L, F. Le Menn, M. Kentouri, M. Kagara and A. Fostier (2001). Seasonal cycle of gonadal development and plasma levels of vitellogenin of the red porgy, *Pagrus pagrus* (Teleostei, Sparidae). *Mar. Biol.*, 139: 549-559.

Koundal A., R. Dhanze and I. Sharma (2014). Length-weight relationship, condition factor and relative growth patterns of *Channa punctata* (Bloch) from Himachal Pradesh, India. *Zoo's Print*, 23 (1): 25-29.

Król J, and E. Zieliński (2015). Effects of stocking density and weaning age on cannibalism, survival and growth in European perch *Perca fluviatilis* larvae. *Pol. J. Nat. Sci.*, 30: 403-415.

Krol Jaroslaw and Zyzislaw Zakes (2016). Effect of dietary L-tryptophan on cannibalism, survival and growth in pikeperch *Sander lucioperca* (L.) post larvae. *Aquacult. Int.*, 24: 441-451.

Kumar D., K. Marimuthu, M.A. Haniffa and T.A. Sethuramalingam (2008). Effect of different live feed on growth and survival of striped murrel *Channa striatus* larvae *E.U. J. Fish. Aquatic Sci*, 25: 105-110.

Kumari Sangeeta, V. K. Tiwari1, A. M. Babitha Rani, Rajesh Kumar and Satya Praksah (2018). Effect of feeding rate on growth, survival and cannibalism in striped murrel, *Channa striata* (Bloch, 1793) fingerlings. *J. Exp. Zool. India*, 21 (1): 205-210.

LARReC (2002). A Report on Cage Culture in Lao PDR. The Living Aquatic Resources Research Center (LARReC) and Network of Aquaculture Centers in Asia Pacific (NACA).

Leu M.Y., Ko-Yuan Tai, Pei-Jie Meng, Cheng-Hao Tang and Pin-han Wang (2018). Embryonic, larval and juvenile development of the longfin batfish, *Platax teira* (Forssskal, 1775) under controlled conditions with special regard to mitigate cannibalism for larviculture. *Aquaculture*, 493: 204-213.

Liao, I. Chiu, Meei Su Huei and Chang Y. Emily (2001). Techniques of finfish larviculture in Taiwan. *Aquaculture*, 200 (1): 1-31.

Lupatsch, I., G.W. Kissil and D. Sklan (2003). Comparison of energy and protein efficiency among three fish species gilthead sea bream (*Sparus auratus*), European seabass (*Dicentrachus labrax*) and white grouper (*Epinephelus aeneus*): energy expenditure for protein and lipid deposition. *Aquaculture*, 225: 175-189.

Mahmod, Nasim Al, Md. Hafizur Rahman Hassan, Golam Mohammod Mostakin, Modh. Golam quadar Khan, Md. Shahjahan, Nahid Sultana Lucky and M. Sadiqul Islam (2016). Cyclic variations of gonad development of an air-breathing fish, *Channa striatus* in the lentic and lotic environments. *Fisheries and Aquatic Sciences*, 19: 1-7.

Maniaca A. (2002). Filial cannibalism in teleost fish. *Biol. Rev. Camb. Philos. Soc.*,77 (2): 261-77.

Marimuthu K. and M. A. Haniffa (2007). Embryonic and larval development of the striped murrel *Channa striatus*. *Taiwania,* 52 (1): 84-92.

Marimuthu K., D. Kumar and M. A. Haniffa (2007). Induced spawning of striped murrel *Channa striatus*, using Ovatide. *J. App. Aquacul*., 19 (4): 95-103.

Marshall C.T., N.A. Yaragina, Y. Lamber and O.S. Kjesbu (1999). Total lipid energy as a proxy for total egg production by fish stock. *Nature*, 402: 288-290.

Mehraj-Uddin W., K. Altaff and M.A. Haniffa (2009). Study on the growth and survival of *Channa striatus* (Bloch) post larvae using live feed. *Bangladesh J. Fish. Res*., 13(2): 131-135.

Mélard C., E. Baras, L. Mary and P. Kestemont (1996). Relationships between stocking density, growth, cannibalism and survival rate in intensively cultured larvae and juveniles of perch (*Perca fluviatilis*). *Ann. Zool. Fennici,* 33: 643-651.

Mohanty S.S. and K. Samantaray (1996). Effect of varying levels of dietary protein on the growth performance and feed conversion efficiency of murrel *Channa striata* fry. *Aquacult. Nutr.*, 2: 89-94.

Mollah M.F.A., M.S.A. Mamum, M.N. Sarower and A. Roy (2009). Effects of stocking density on the growth and breeding performance of broodfish and larval growth and survival of shol, *Channa striatus* (Bloch). *J. Bangladesh Agril. Univ*. 7 (2): 427-432.

Morioka Shinsuke, Bounsong Vongvichith, Phonaphet Chanthasone, Phoutsamone Phommachane and Naoki Suzuki (2016). Reproductive season, age estimation and growth in a striped murrel *Channa striata* population in Nasaythong District, Vientiane Province, Central Laos. *Aquacult. Sci*., 64 (2): 183-191.

Murugesan V. K. and P. Kumaraih (1976). Progress report of the work done by the Karnataka center. *In:* Third workshop of the All India Coordinated Research Project on Air breathing Fish Culture, CIFRI, Barrackpore, India, 29pp.

Muthumainnah Dina (2013). Growout of striped murrel (*Channa Striata*) in swamp water system using fences and cages. 4th International Conference on Biology, Environment and Chemistry IPCBEE vol. 58, IACSIT Press, Singapore, pp 52-55.

Narejo N.T., S. Jalbanil , and G. Dastagir (2015). Breeding Biology of Murrel, *Channa striatus* (Bloch) from District Badin Sindh, Pakistan. *Biolife*, 3 (2): 434-436.

Nen phanna, Phen Chheng, Nam So, Tran Thi Thanh Hien, Bui Minh Tam, Hillary Egna and David A. Bengtson (2018). Performance of domesticated (Vietnamese) versus non-domesticated (Cambodian) murrel, *Channa striata* (Bloch 1793) with regard to weaning onto pellet feed. *Asian Fishe. Sci.*, 31: 209-217.

Newhard J. Joshua (2015). Identification and location of testes in the invasive *Channa argus* Cantor (Northern murrel). *Northeastern Naturalist*, 22 (3): 16-17.

Ng P.K. and K.K.P. Lim (1990). Murrels (Pisces: Channidae): Natural history, biology and economic importance. *In: C.L. Ming and P.K.L. Ng (Editors), Essays in Zoology*. Papers commemorating the 40th Anniversary of the Department of Zoology, National University of Singapore, pp.127-152.

Nguyen Huan and Duong Nhut Long (2008). The hatchery status and technical aspects for murrel spawning (*Channa micropeltes*). Aquaculture and Fisheries College, Can Tho University, Vietnam (Tap chi Khao hoc (2): 20-28.

Olurin K. B. and O. D. Savage (2011). Reproductive biology, length-weight relationship and condition factor of the African murrel, *Parachanna obscura*, from River Oshun, South-west Nigeria. *Inter. J. Fishe. Aquacul.* 3 (8): 146-150.

Ortega V.A., R.J. Renner and N.J. Bernier (2005). Appetite-suppressing effects of ammonia exposure in rainbow trout associated with regional and temporal activation of brain monoaminergic and CRF systems. *J. Exp. Biol.*, 208: 1855-1866.

Parameswaran S. and V. K. Murugesan (1976a). Breeding season and seed resources of murrels in swamps of Karnatka state. *J. Indian. Fish. Soc. India*, 8: 60-67.

Parameswaran S. and V. K. Murugesan. (1976b). Observation on the hypophysation of murrels (Ophiocephalidae). *Hydrobiol.*, 50 (1): 81-87.

Pati M.K., S.C. Rath and P.C. Thomas (2004). Brood rearing, induced spawning and egg incubation of *Channa striatus* in an indoor system with a note to its second breeding in the same season. *J. Aqua.* 12: 7-12.

Pedro N.D., M.L. Pinillos, A.I. Valenciano, M. AlonsoBedate and M.J. Delgado (1998). Inhibitory Effect of Serotonin on Feeding Behavior in Goldfish: Involvement of CRF. *Peptides*, 19: 505-511.

Pethiyagoda P. (1991). Freshwater fishes of Sri Lanka: Colombo Wildlife Heritage Trust of Sri Lanka, 362p.

Qasim S. Z. and A. Qayyum (1961). Spawning frequencies and breeding season of some freshwater fishes with special reference to those occurring in the plain of Northern India. *Ind. J. Fishe.*, 8: 24-43.

Qin Jianguang and A.W. Fast (1996). Size and feed dependent cannibalism with juvenile murrel *Channa striatus*. *Aquaculture*, 144: 313-320.

Qin J., X. He and A. W. Fast (1997a). A bioenergetics model for an air breathing fish *Channa striatus*. *Environ. Biol. Fishes*, 50: 309-318.

Qin J.G. and A.W. Fast (1997b). Food selection and growth of young murrel (*Channa striatus*). *J. Applied Ichthyology*, 13: 21-26.

Qin JG and A.W. Fast (1998). Effects of temperature, size and density on culture performance of murrel, *Channa striatus* (Bloch), fed formulated feed. *Aqua. Res.*, 29: 299-302.

Qin J.G., L. Mittiga, and F. Ottolenghi (2004). Cannibalism reduction in juvenile barramundi *Lates calcarifer* by providing refuges and dim light. *J. World Aquacul. Soc.*, 35: 113-118.

Qinfen Dong, Peng Zhidong, Mei Gaoshang, Zhang song and Yang Yong (2013). Hybrid murrel farming in China. *Aqua Culture Asia Pacific Magazine*, Nov-Dec., p32-36.

Rahman A.K.A. (1989). Freshwater fishes of Bangladesh. Zoological Society of Bangladesh. Department of Zoology, University of Dhaka, 364p.

Rahman M. Aminur and Sadiqul Awal (2016). Development of captive breeding, seed Production and culture techniques of murrel fish for species conservation and sustainable aquaculture. *Int'l J. Adv. Agricultl. Environmental Engg.*, 3 (1): 117-120.

Rahman M. A., A. Arshad and S.M.N. Amin (2012). Growth and production performance of threatened murrel fish, *Channa striatus* (Bloch), at different stocking densities in earthen ponds. *Aquacul. Res.*, 43 (2): 297-302.

Rahman, M.A., A. Arshad, S.M.N. Amin and Mariana Nor Shamsudin (2013). Growth and survival of fingerlings of a threatened murrel, *Channa striatus* (Bloch) in earthen nursery ponds. *Asian J. Anim. Vet. Adv.*, 8 (2): 216-226.

Raizada S., P.P. Srivastva, P. Punia, K. C. Yadav, V. Sahu, S. Chaudhary and J.K. Jena (2012). Dietary Protein requirement of Giant Murrel, *Channa marulius* (Ham., 1822) fry and impact on growth indices. *Natl. Acad. Sci., India, Sect. B. Biol. Sci.*, 82 (4): 489-496.

Raizada Sudhir, Prem Prakash Srivastava, Vikash Sahu, Kailash Chandra Yadav and J. K. Jena (2015). Observations on captive breeding of the threatened freshwater shark *Wallago attu* (Bloch & Schneider, 1801). *Indian J. Fish.,* 62(4): 120-124.

Rao T. and C.V. Narasimha (2020). Experience in farming the murrel in India. *Aqua Culture Asia Pacific Magazine*. May-June, 2020, 40-41p.

Rawat Anurag, Sudhir Raizada, Kailash Chandra Yadav, Vikash Sahu (2020). Induced spawning and embryonic development of striped murrel, *Channa striatus* (Bloch, 1793) in indoor conditions with sGnRH analogue. *Indian J. Fishe.*, 76 (3): 52-61.

Ribeiro F.F. and J.G. Qin (2015). Prey size selection and cannibalistic behavior of juvenile barramundi *Lates calcarifer*. *J. Fish Biol.*, 86 (5): 1549-1566.

Rohwer S. (1978). Parental cannibalism of offspring and egg raiding as a courtship strategy. *American Naturalist*, 112: 429-440.

Roy Nirmal Chandra, Sajib Kumar Chowdhury and Subrata Kumar Das (2016). Observation of hapa breeding technique of striped murrel, *Channa striatus* (Bloch, 1793) under captive conditions, *Internatl. J. Fishe. Aquat. Stud.*, 4 (5): 413-417.

Samantaray K. and S.S. Mohanty (1997). Interactions of dietary levels of protein and energy on fingerling murrel, *Channa striata. Aquaculture*, 156 (3-4): 241-249.

Sangavi, S., Alok Kumar Jena, Sandeep Shankar Pattanaik and C. Judith Betsy (2019). Effect of dietary l-tryptophan supplementation on growth and survival of striped murrel, *Channa striatus* (Bloch, 1793) fry. *Global J. Bioscie. Biotechnol.*, 8 (3): 307-311.

Saputra Adang, Tatag Budiardi, Reza Samsudin and Naufal Dwi Rahmadya (2018). Growth performance and survival of murrel *Channa striata* juvenile with different stocking density reared in recirculation system. *Jurnal Akuakultur Indonesia* 17 (2): 104-112.

Sargent R.C. (1992). Ecology of filial cannibalism in fish: theoretical perspectives. *In: Cannibalism: Ecology and Evolution Among Diverse Taxa* (Ed. M. A. Elgar & B.J. Crespi), Oxford: Oxford University Press, 38-62pp.

Sarower M.N., M.Z.H. Jewel, M.A. Sayeed and M.F.A. Mollah (2010). Impacts of different diets on growth and survival of *Channa striatus* fry. *Int. J. Bio. Res.* 1 (3): 8-12.

Selvaraj S. and T. Francis (2007). Influence of human chorionic gonadotropin on maturation in striped murrel, *Channa striatus*. *Asian Fishe. Sci.* 20: 23-39.

Selvaraj, S., T. Francis, M. Venkatasamy and S. Santhoshkumar (2012). Effect of HCG implants on change in testosterone and estradiol level in blood serum of murrel, *Channa striatus*. *Tamilnadu J. Vet. Anim. Sci.*, 8 (5): 290-298.

Smith C. and P. Reay (1991). Cannibalism in teleost fishes. *Rev. Fish Biol. Fishe.*, 1:41-64.

So Nam, Robert S. Pomeroy and Dong Thap (2012). Sustainable murrel aquaculture development in the lower Mekong River Basin of Cambodia and Vietnam Indigenous Species Development / Study /09IND02UC Part 1: Breeding and weaning of striped murrel (*Channa striata*) in Cambodia. Corpus ID: 53954029.

Sogard S.M. and B.L. Olla (1994) The potential for intracohort cannibalism in age-0 walleye pollock, *Theragra chalcogramma*, as determined under laboratory conditions. *Environ Biol Fishes,* 39: 183-190.

Stiassny, M. L. J. and N. Raminosoa (1994). The fishes of the inland waters of Madagascar. In "Biological diversity of African fresh- and brackish water fishes" (ed. by G. G. Teugels, J.-F. Guégan and J.-J. Albaret). *Ann. Mus. R. Afr. Centr., Sci.* Zool.,275: 133-148.

Subagja Jojo, Jacques Slembrouck, Le Thanh Hung and Marc Legendre (1999). Larval rearing of an Asian catfish *Pangasius hypophthalmus* (Siluroidei, Pangasiidae): Analysis of precocious mortality and proposition of appropriate treatments. *Aquat. Living Resour.*, 1 2 (I): 37-44.

128. Szczepkowski M. (2009). Impact of selected abiotic and biotic factors on the results of rearing juvenile stages of northern pike *Esox lucius* L. in recirculating systems. *Arch. Pol. Fish.*, 48: 107-147.

Thumronk Amornsakun, Wasan Sriwatana and Ponpanom Promkaew (2011). Some aspects in early life stage of murrel fish, *Channa striatus* larvae. *Songklanakarin J. Sci. Technol.* 33 (6): 671-677.

Thumronk Amornsakun, Vuong Hung Vo, Nirattisai Petchsupa, Tan Min Pau, and Anuar bin Hassan (2017). Effects of water salinity on hatching of egg, growth and survival of larvae and fingerlings of murrel fish, *Channa striatus*. *Songklanakarin J. Sci. Technol.* 39 (2): 137-142.

Tyler C., J. Nagler, T. Pottinger and M. Turner (1994). Effects of unilateral ovariectomy on recruitment and growth of follicles in rainbow trout, *Oncorhynchus mykiss*. *Fish Physiol. and Biochem*, 13: 309-316.

Wan A. Aziz, Yakob Wan and B. Ali Ahyaudin (1992). Simple method for backyard production of murrel (*Channa striatus* Bloch) fry. *Naga The ICLARM Quarterly* April 22-23.

War M., A. Kareem and M. A. Haniffa (2011). Growth and survival of larval murrel *Channa striatus* (Bloch, 1793) fed different live organisms. *Turkish J. Fishe. Aquat. Sci.*, 11: 523-528.

Wee K.L. (1981). Murrel (*Channa striatus*) farming in Thailand. NACA/WP/81/3 Nov., 1981..

Welcomme, R. L. (1988) International introductions of inland aquatic species. FAO Fish. Tech. Pap., 294, Rome, 318p.

Willey A. (1909). Observation on the nests, egg and larvae of *ophiocephalus striatus*. *Spoliya Zeylanica*, 6: 108-123.

Yaakob W.A.A. and A.B. Ali (1992). Simple method for backyard production of murrel (*Channa striata,* Bloch) fry. *Naga*, 15 (2): 22-23.

Yadav K.C., A. Mishra, S. Raizada, V. Sahu and P. P. Srivastava (2014). Influence of formulated diet on survival and growth of giant-murrel, *Channa marulius* reared in pond condition. *Proc. Natl. Acad. Sci., India, Sect. B Biol. Sci.* 86(1): 97-103.

Yadav Kailash Chandra (2016). Study on reproductive biology, captive breeding, larval rearing and grow-out of giant murrel, C*hanna marulius* (Hamilton, 1822). Ph.D Thesis. Baba Saheb Bhimrao Ambedkar, University (Central University), Lucknow-226025, India.

Yadav Kailash Chandra, Sudhir Raizada and Abha Mishra (2016). Study of gonado- somatic index of male and female giant murrel fish, *Channa marulius* (Hamilton, 1822). *J. Exp. Zool. India*, 19 (2): 719-721.

Yulintine Hendri Bugar, Linda Wulandari and Edison Harteman (2017). Murrel Fish (*Channa Striata*): Semi-Induced Breeding and Larval Growth. *Indian J. Sci. Technol.*, 10 (11): 1-8.

Appendix

I. Methodology of Brine Shrimp (Artemia) Nauplii Production

Artemia is a small microscopic animalcule of crustacean group which develop in salt pans and high saline water bodies. During unfavourable climate, they breed and form hard shelled cyst (eggs). These cyst hatch on getting suitable climate and give birth to nauplii, a free swimming larvae. They form a highly nutritious live diet for the larvae of prawn, however their nauplii size is important for proper use in the hatchery. *Artemia salina* of Great Salt Lake (USA) is comparatively very small (~500µm in length) and easily consumed by fish and prawn larvae. Artemia nauplii are sold in the market in tin packs of generally one pound weight.

Artemia cyst are by nature contaminated with bacteria, fungal spores, other micro- organisms and organic impurities that may infect the water of larval rearing tanks, if not properly treated. In order to avoid such situation, cysts need to be disinfected/ decapsulated. The disinfection eliminates the chances of infection; whereas decapsulation process eliminates infection better and also improves the energy content of artemia nauplii.

In order to get nauplii of brine shrimp, the following steps are to be followed.

Step-1: Estimation of quantity of BSN

- Calculate the quantity of nauplii needed for feeding to larvae.
- The number of artemia per unit weight depends on the strain of artemia. On an average, the artemia of Great Salt Lake (USA) may yield 2.7 lakh and that of San Francisco Bay (USA) 3.2 lakh per gram weight.
- Find out the hatchability rate of artemia from the instructions written over the box for first time and subsequently after observing the hatching percentage by manual counting.
- Calculate the number of cyst required by adjusting hatching rate.
- Find out the weight of artemia cyst required for getting the calculative quantity of nauplii.

Step-2: Hydration of cysts

- Weight the desired quantity of artemia cyst in a balance.
- Take seawater/ inland saline water/ freshwater in a container (preferably with conical bottom) and switch on aeration. The quantity of water may be sufficient to keep the eggs moving during aeration. Small quantities may need more water. A litre of water for 10-100gm cyst is sufficient.
- Keep cyst in the container for about an hour for hydration.
- Filter the cysts into a scoop net of 100 micron mesh bolting silk cloth.

Step-3: Disinfection of Brine Shrimp Cyst

- Disinfection of artemia cyst can be done by immersing them in a hypochlorite solution (10-12% or 200 ppm active chlorine)/ treating with bleaching powder.
- Since the chlorine content of hypochlorite and bleaching powder changes after storing, it is essential to determine the percentage of active chlorine (%Cl) in the commercial sodium hypochlorite/ bleaching powder.

Step-4: Decapsulation of cysts

A. Decapsulation with sodium hypochlorite (NaOCl)

- Take hydrated cysts in a beaker. Cyst on dehydration changes from concave appearance to spherical in shape.
- Cyst are then drained through the 125 μm screen bolting silk cloth and transferred to a bucket with seawater.

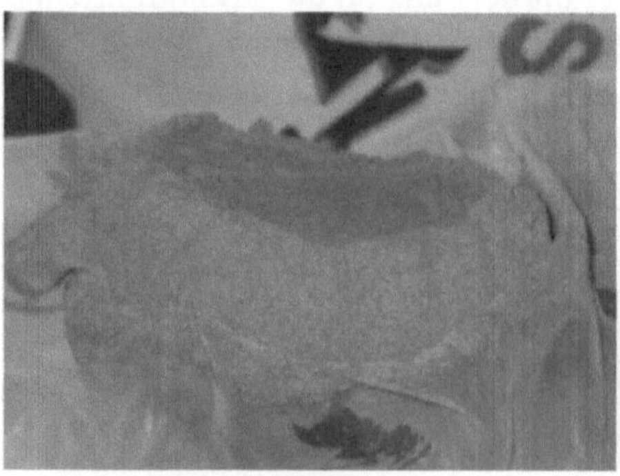

Fig. 54: Soaked artemia cyst

Fig. 55: Preparation of hypochlorite solution

- The sodium hypochlorite/sodium hydroxide solution is added and the mixture constantly stirred with the glass tube.

- Keep this beaker in trough containing cold water/ iced water so that the temperature of beaker water should not rise (reach beyond 40 ^{0}C) to damage the artemia embryos.
- For decapsulation of 100gm of cyst, cysts are to be put in 900ml seawater and 500ml sodium hypochlorite + 15gm NaOH is added.
- Note the change of colour of the cyst from dark brown to white and then orange resulting due to dissolving of chorion layer of cyst. This generally occurs within 5 to 10 minutes.
- Filter the cysts immediately on a 100 micron mesh bolting silk cloth.

Fig. 56: Decapsulated artemia cysts

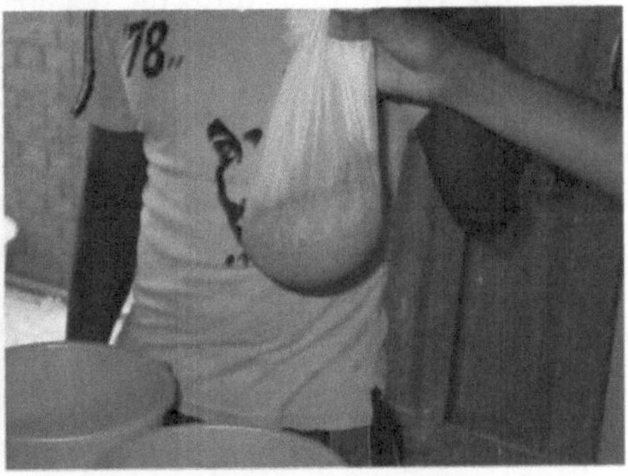

Fig. 57: Washing of decapsulated cysts

- Clean thoroughly the decapsulated cysts in water (seawater/ salinewater/freshwater) to avoid traces of toxic chlorine.
- A dip treatement with 0.1% sodium thiosulphate solution may be given for complete removal of residual chlorine.

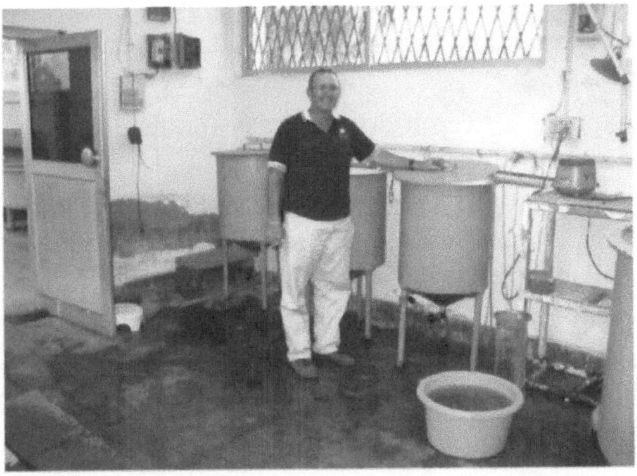

Fig. 58: Artemia tanks and collection of nauplii in tub

B. Decapsulation with bleaching powder

- Take dehydrated cyst. Add lime (CaO) @ 12.5% of the weight of eggs, Stir until mixed thoroughly. Add bleaching powder @ 27.5% of the weight of artemia and keep on stirring until the substances are mixed thoroughly. The procedure must take place at a controlled temperature of 40°C. Stir for 5-8 minutes until the temperature stays steady.
- Add some more ice to decrease the temperature to 30oC and then continue with a second decapsulation stage in the same manner as the first stage, adding lime and bleaching powder in the same manner. Stir for 5-8 minutes until the decapsulation reaction is complete. The colour of the eggs will change from dark brown to white. Leave them at that stage until the white colour gradually becomes orange. The whole procedure, including both stages takes 10-16 minutes.
- Transfer the eggs into a filter bag. Wash them with several changes of clean water to remove the smell of chlorine and presence of dissolved chemicals
- Put the eggs in the solution of sodium thiosulphate prepared to provide strength of not more than 0.05gm per 100gm artemia cysts.
- Add 100ml of waterm, stir for 2-5 minutes and mixed thoroughly. The decapsulated eggs will sink to the bottom of the container.

Step-4: Hatching of decapsulated cysts

- This process is carried out in long glass jar (for small quantity) or cylindro- conical FRP tank (for large quantity) having translucent bottom (capacity 10- 500 litre).
- Clean the jar/ tank with water and disinfect with 200mgL^{-1} chlorine or 100mgL^{-1} formalein.
- Wash thoroughly the tank and keep it dry for a day (preferable in day light) so that residual chlorine removes.
- Fill the jar/ tank with 25-30gL^{-1} salinity seawater. Seawater, inland saline water and freshwater can be used and salinity may be adjusted by adding common salt and freshwater in case of dilution.
- Switch on aeration from a blower at the bottom of the container. The concentration of oxygen should not drop less than 2mgL^{-1}.
- Arrange florescent light on top of the container to get light of 2000 lux intensity (two tube lights of 40 watt). Maintain a temperature of 25-30°C of water.

- Transfer the cysts in the container.
- Cysts hatch out in about 12-24 hours depending on the strain of artemia.

Step-5: Harvesting of nauplii

- Artemia nauplii attract towards light, this behaviour of nauplii is used for the collection of nauplii from the artemia tanks.
- First of all, stop aeration in the tank.
- Wait for water till swinging action is stopped. With this all sediments and cysts both hatched and unhatched will settle at the bottom of the tank. Open the outlet at the bottom of the tank to drain sediment. The nauplii will remain floating at the surface of the tank due to illumination of florescent lamp. Close the drain valve.
- Switch off the florescent light at the top of the tank and cover the mouth with a lid or black cloth. Now, illuminate a lamp near the translucent conical part of tank to concentrate artemia at the bottom and when all the artemia come near the bottom, drain out in a conical bag net made up of 100 micron bolting silk kept in a bucket filled with water and having an outlet near the top edge for flowing out excess of water.
- In case of not having above type of jar, the artemia hatching may be undertaken in any other container. The aeration may be stopped after hatching of most of the cyst. Cover the entire container with black cloth except the top of the container. Put a lamp at the top that will attract artemia napulii at the top fom where they are siphoned type with tubing.
- Wash the nauplii thoroughly with water of same salinity as that of artemia hatching tank.

Fig. 59: Freshly hatched BSN

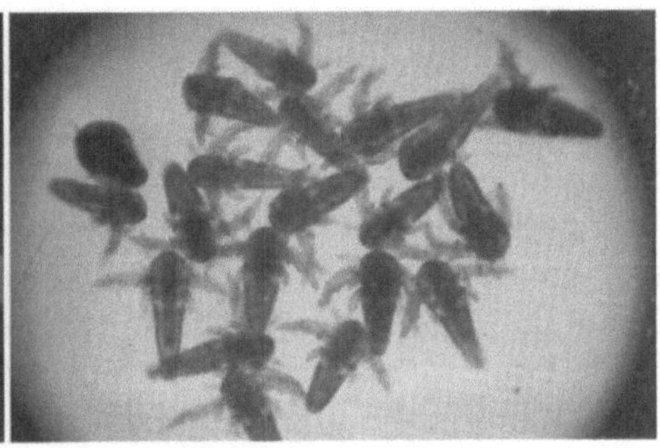

Fig. 60: BSN under microscope

Step-6: Stocking of nauplii

- Keep all nauplii in a bucket of water having same salinity as that of larval rearing water.
- Count the number of nauplii in one millilitre of water and transfer the required quantity of nauplii along with water in each tank.

Step-7: Preservation of eggs

- The decapsulated eggs may be stored in the refined salt in the ratio of 30gm NaCl/100gm of brine shrimp eggs. The salt dehydrates the eggs and they can be stored in fridge for later use.

Precautions

- Artemia cyst should be kept in refrigerator as it lost hatchability as well as nutritional quality at higher temperature.
- During encapsulation process, the temperature of water should not exceed above 40°C.
- As soon as the cyst turns orange during decapsulation process, they should be removed from the decapsulation solution and washed with sea or freshwater, otherwise the artemia embryos may damage.
- Arrange vigorous aeration at the bottom of the tank so as to avoid settling of cysts.

Calculation for Hatching Percentage

Hatching (%) = No. of nauplii x 100 / No. of cysts + nauplii

II Artemia Enrichment

The nutritional quality and physical size of nauplii vary enormously from source to source and even between individual batches from a single source. Of particular importance is the level of essential polyunsaturated fatty acids, eicosapentaenoic acid (EPA) (20:5n-3) and docosahexaenoic acid (DHA, 22:6n-3), which depends on the composition of the primary food available to the brine shrimp in the locations, where they originate and is generally found low. In order to provide sufficient quantity of these essential fatty acids, the nauplii are to be enriched with both EPA and DHA. There are various enrichment products available in the market like Super Selco, DHA Selco (INVE, Aquaculture), Super artemia (Catvis BV., 5222 AE, Netherland), Super HUFA (Salt Creek Inc, USA), Algamac-2000, Algamac-3050 (Biomarine Inc.,

USA). The methodology for enrichment is provided with these products. The methodology of using Algamac – 2000 (used by the authors) is as under.

Measure 30gm of Algamac-2000 and blend in 1.0 litre water for about one minute. Put the BSN in this media for about 1-2 hours with vigorous aeration and harvest for feeding to larvae.

III *Moina* Culture

Moina spp. is a crustacean microscopic organism commonly found in freshwater ponds/ tanks and lakes. It multiplies through parthenogenesis under normal conditions. An adult can produce 10-12 nos. per day the young ones become adult in 18-24 hours and begin to reproduce. It is possible to obtain 42,000 nos. within 12 days from a single female.

Mass culture of *Moina* may be undertaken in FRP and Cement Tanks of size 1-10 tonne. The tanks to be filled up with freshwater and fertilized with groundnut oil cake (250g), urea (10g) and single super phosphate (25g) per tonne of water. The tank are then inoculated with freshwater Chlorella sp. same day with a cell concentration of 30-50 million cell per ml with 5-10 litre of culture water per tonne. The tank water to be aerated vigorously. On second day when the tank water turns light green, *Moina* @ 1-5 nos. per litre of water, to be inoculated from the stock solution. *Moina* will start feeding on the Chlorella and fine particles of groundnut oil cake and within 5-7 days, *Moina* concentration of 25,000-30,000 will reach. Now, 30-50 percent of *Moina* population can be harvested. The harvested water to be replenished with freshwater and also replenished with proportionate amount of fertilizers used above as well with Chlorella. *Moina* can be harvested in the morning and evening with the help of a plankton net. The harvested *Moina* has to be thoroughly washed with freshwater before feeding to the larvae. Since *Moina* is a freshwater organism, it is poor in Omega-3 PUFA. It is therefore better to enrich *Moina* before it is given as feed to the larvae.

Colour Plates

Chapter 2: Reproductive Biology

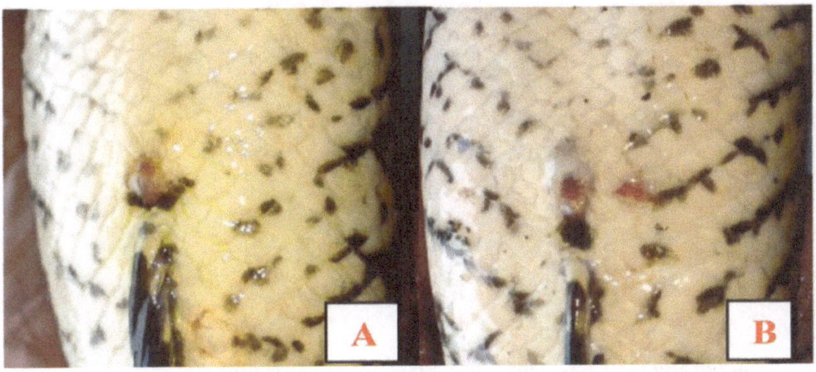

Fig. 2: (A) Gonopore of female with swollen vent and no space between vent and anal fin. **(B)** Gonopore of male with anal papilla and space between pailla nd anal fin

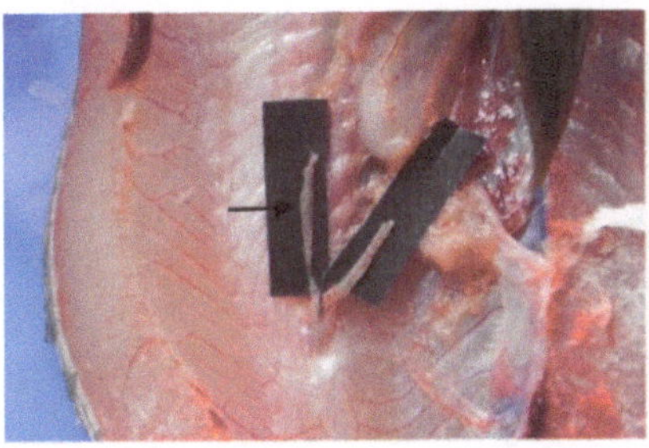

A. Immature Phase of Testes *in situ*

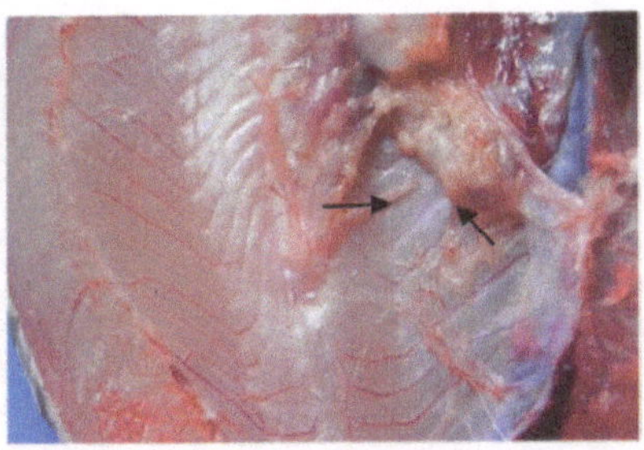

B. Maturing Phase of Testis *in situ*

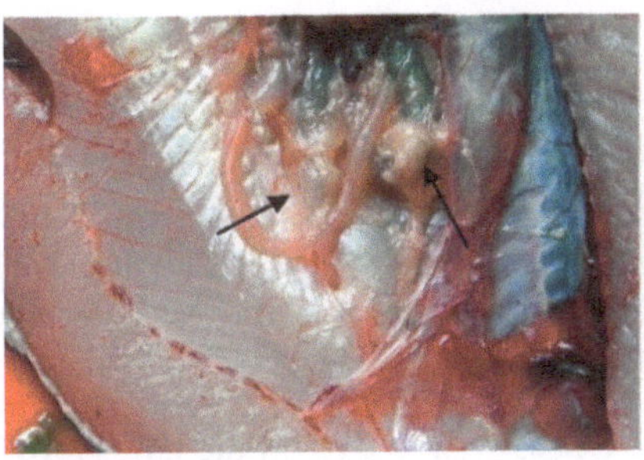

C. Mature Testes *in situ*

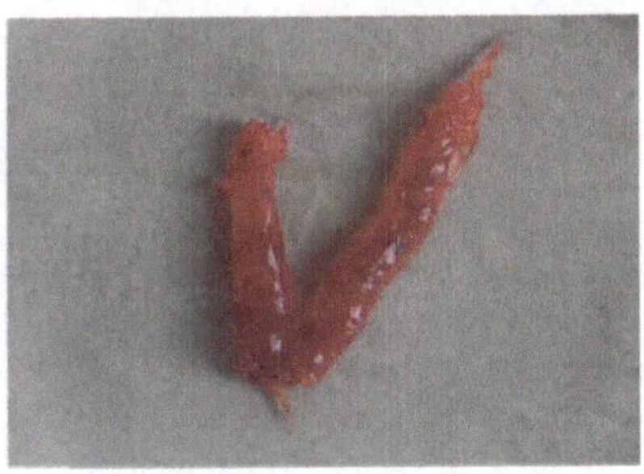

D. Extracted Testes

Fig 3: Developing stages of testes in striped murrel (*Channa striata*)

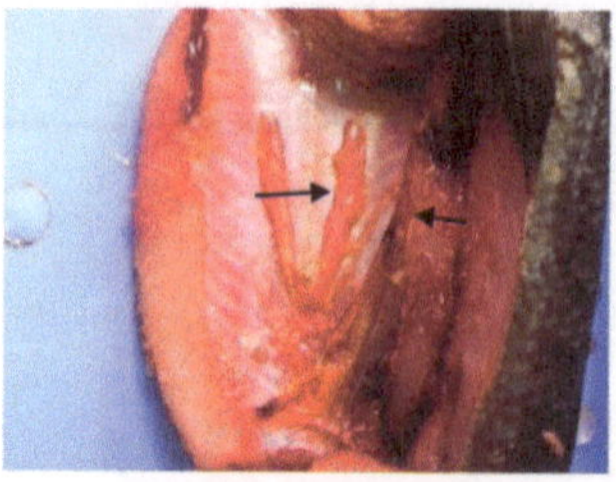

A. Immature Ovary *in situ*

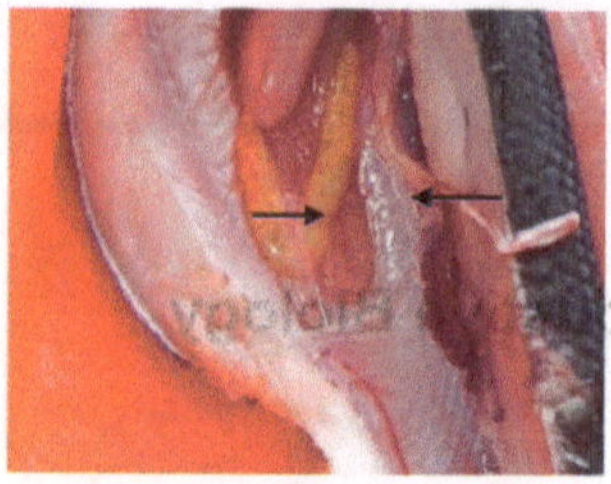

B. Developing Ovary *in situ*

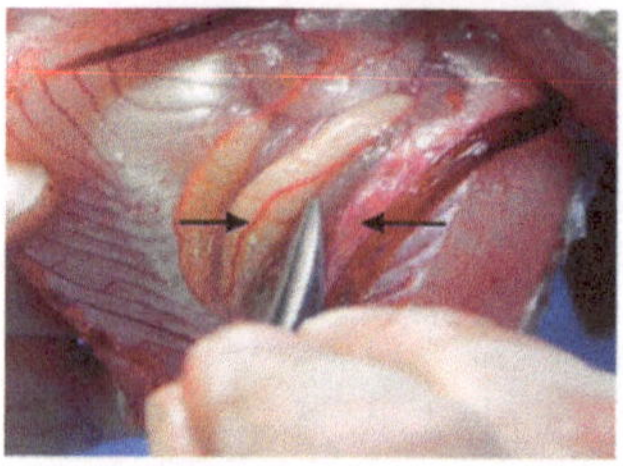

C. Developed Ovary *in situ*

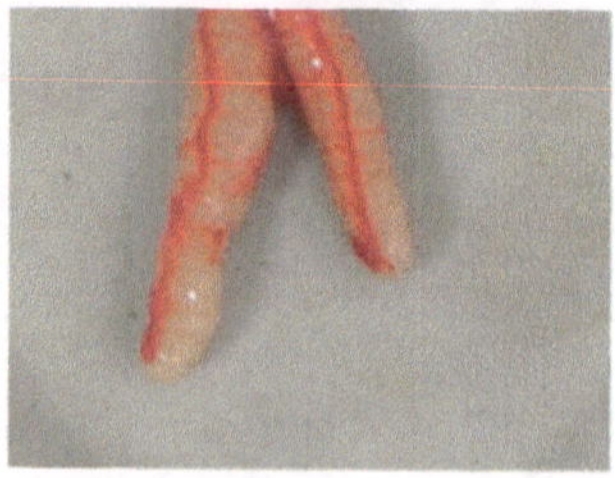

D. Developed Ovary after extraction

Fig. 4: Developing stages of Ovary in striped murrel (*Channa striata*)

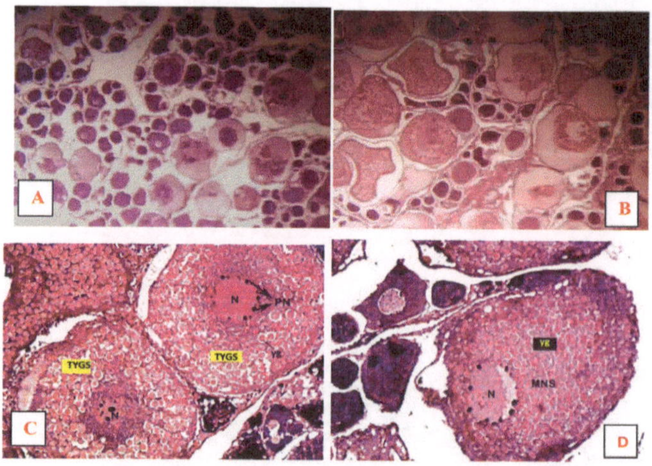

Fig. 5: T.S. of ovary of striped murrel (*C. striata*) **A:** Showing four stages of oocyte development, chromatin nucleolus stage (CNS), early perinucleolus stage (EPS), late perinucleolus stage (LPS), yolk vesicle stage (YVS), indicating immature stage of ovary (Magnification 40x); **B:** Showing four stages of oocyte development, chromatin nucleolus stage (CNS), early perinucleolus stage (EPS), late perinucleolus stage (LPS), yolk vesicle stage (YVS), indicating early maturity stage of ovary (Magnification 40X); **C:** Showing oocyte in the tertiary yolk globular stage showing mature oocyte stage of ovary (Magnification 100X); **D:** Showing migrating nucleus stage (MNS) along with yolk globule oocyte indicating mature stage of ovary (Magnification 100X).

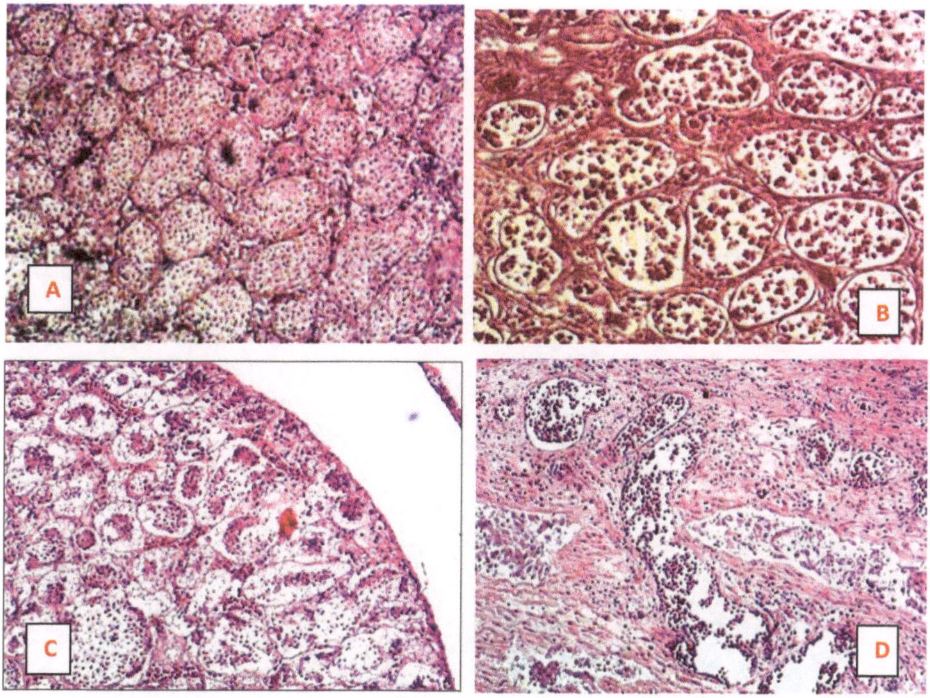

Fig. 6: T.S. of testes of striped murrel (*C. striata*) from cranial part **A:** Showing well organized seminiferous tubules containing spermatocyte in primary and secondary developmental stage (Magnification 40X); **B:** showing advanced stage of spermatogenesis inside the seminiferous tubules, where spermatocytes have been disintegrated from the leydig cells (Magnification 40X); **C:** Cross section depicts Tunica albuginea (outer membrane) of single lobe of testis and spermatozoa in different developmental stages with advance stage moving towards centre of lobule (Magnification 40X); **D:** Showing some seminiferous tubule with spermatozoa, and some empty after release. The vas efference is full of spermatozoa. (Magnification 100X)

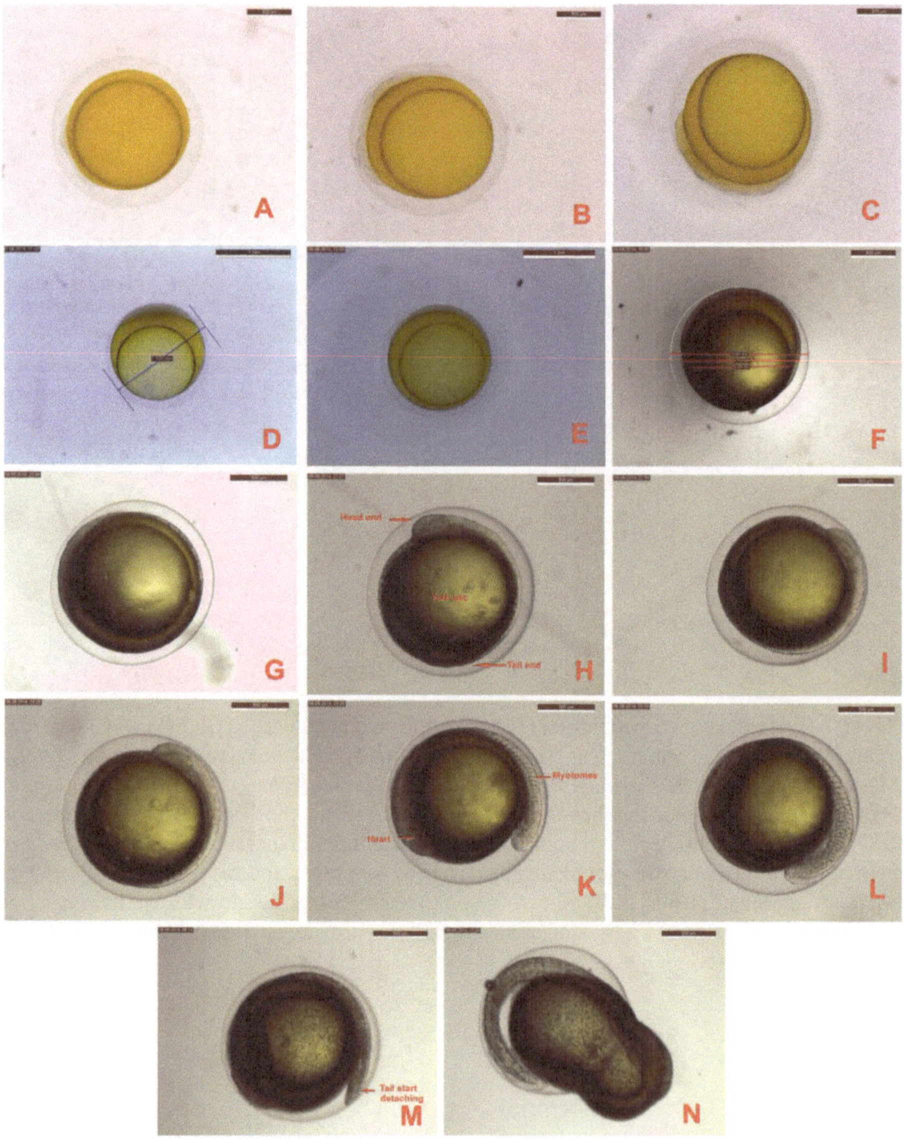

Fig. 7: Developmental stages of egg of striped murrel till hatching

Fig. 8: Developmental stages of hatchling till yolk-sac absorption of striped murrel

Chapter 3: Spawning in Wild

Fig. 9: Nest of striped murrel containing egg mass in pond

Chapter 4: Captive Spawning

Fig. 10: Brooders of striped murrel

Fig. 11: A female with bulging belly, swollen vent and no space between vent and anal fin origin

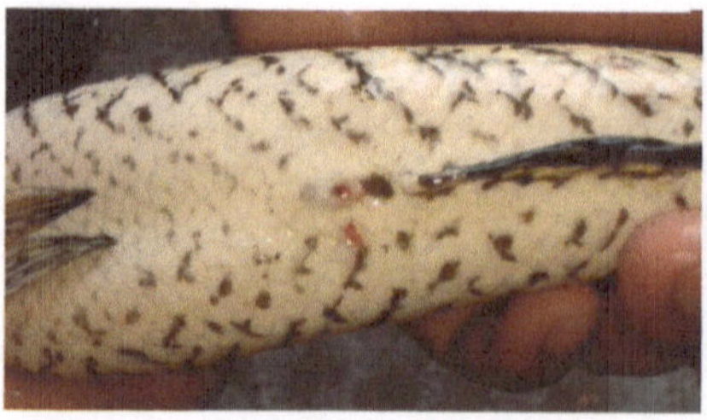

Fig; 12: A male with slender abdomen, spawningun-swollen vent and distinct space between vent and anal fin origin

Fig. 13: Shallow depth FRP tanks for indoor

Fig; 14: Single breeding hapa fixed in pond

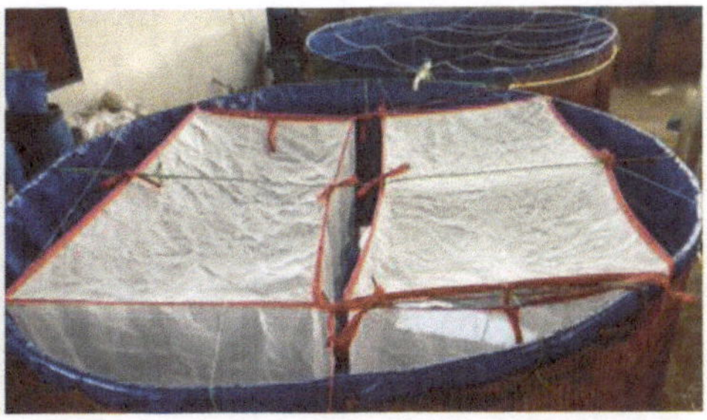

Fig.15: Breeding hapas fixed in indoor lined tank

Fig. 16: Selected two pairs of brooders

Fig. 17: Intra-peritonial injection being given to male fish

Fig. 18: A large egg mass of fertilized eggs in FRP tank

Fig.19: Eggs are being collected from spawning tanks

Fig. 20: Fertilized eggs in plastic mug and a petri dish

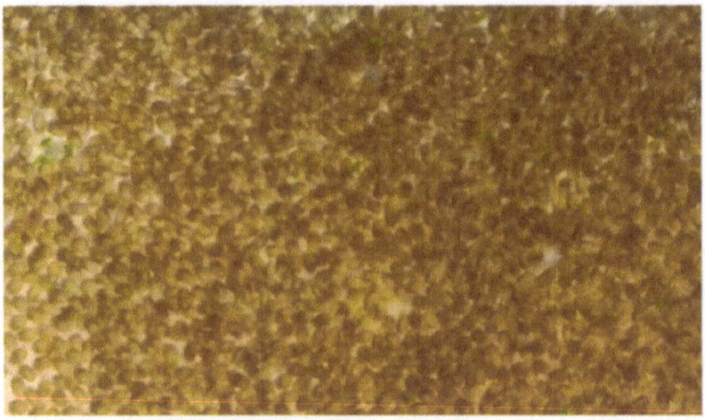

Fig. 21: Hatchling of striped murrel

Chapter 5: Cannibalism and Mitigation

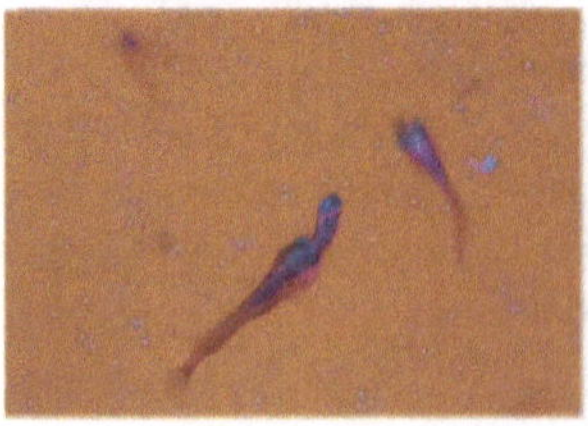

Fig. 22: Predation from tail end

Fig. 23: Predation from head

Fig. 24: Distended abdomen after complete engulfment of the prey

Fig. 25: Manual fish grader of plastic

Fig. 26: Fish size graders made of tin

Chapter 6: Larval Rearing

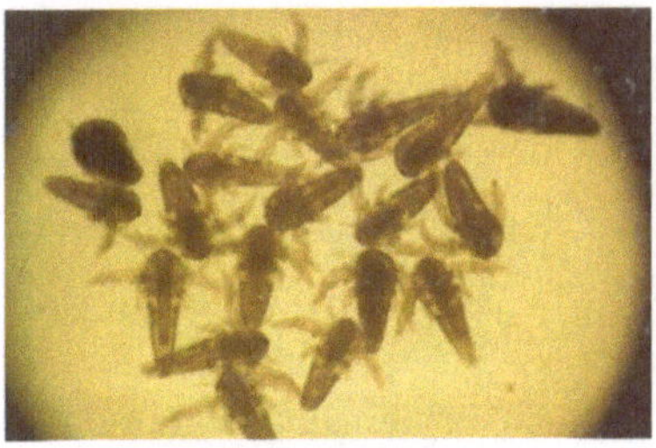

Fig. 27: BSN nauplii under microscope

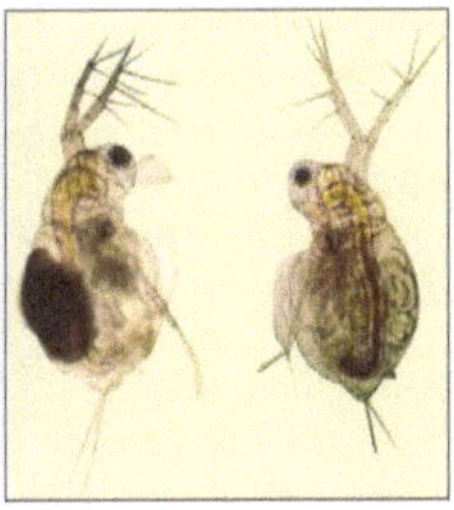

Fig. 28: Moina under microscope

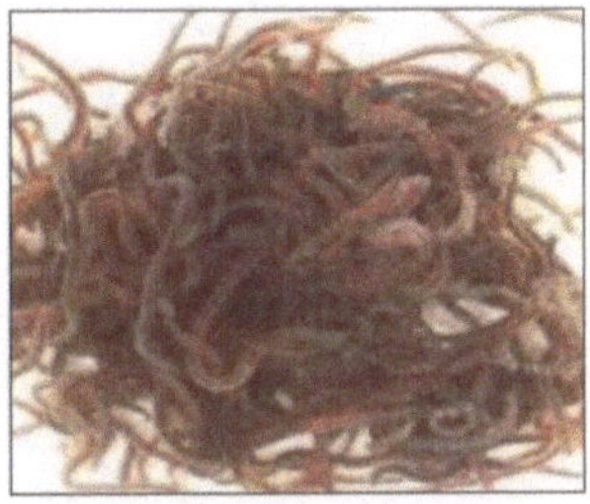

Fig. 29: Tubificid worms

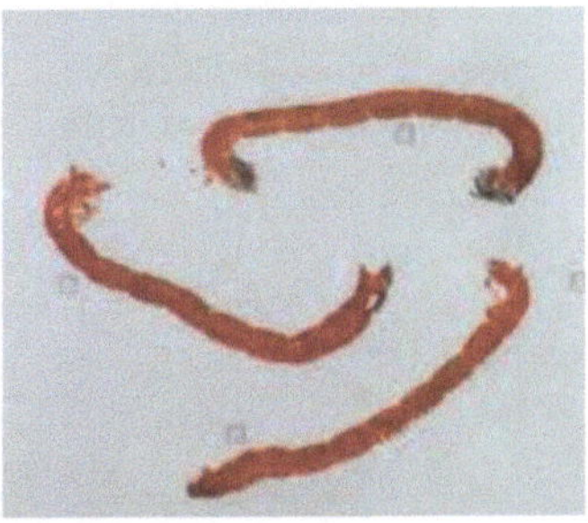

Fig. 30: Blood worms (*Chironomous larvae*)

Fig. 31: Total fish mash

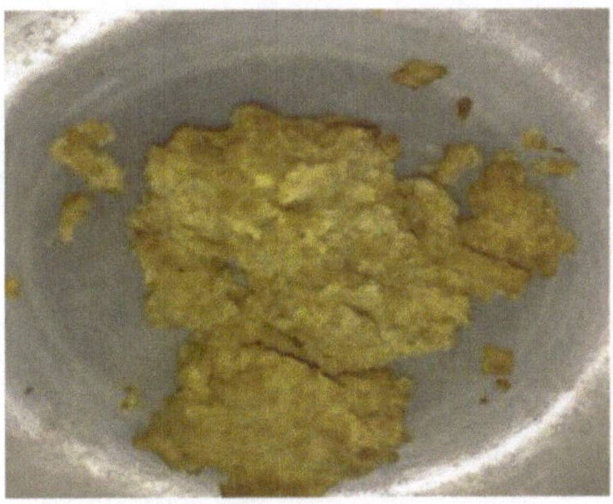

Fig. 32: Fish-mash mixed with grounded formulated diet

Fig. 33: Larval pellet diet

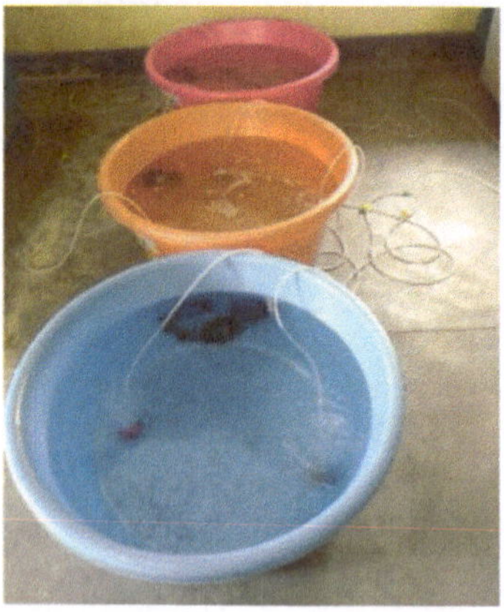

Fig. 34: Incubation of eggs, black co lour due to development of eyes in the eggs

Fig. 35: Early larvae of striped murrel

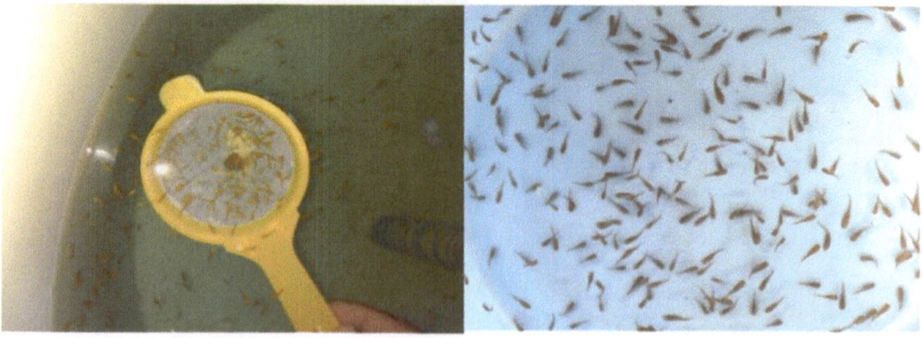

Fig. 36: Rearing of 15 day old larvae in FRP tank

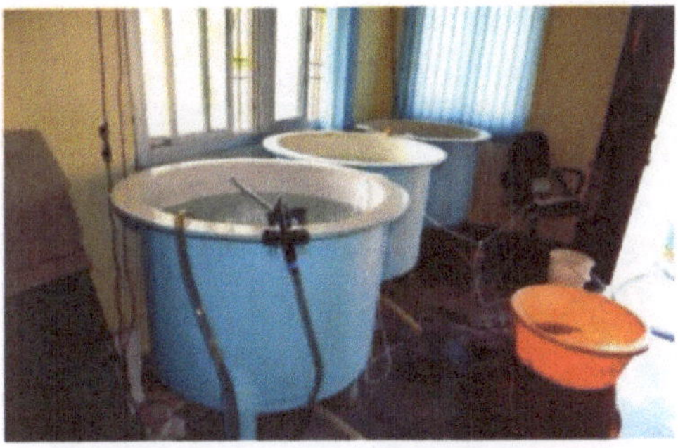

Fig. 37: Larval rearing tanks of 300 litre

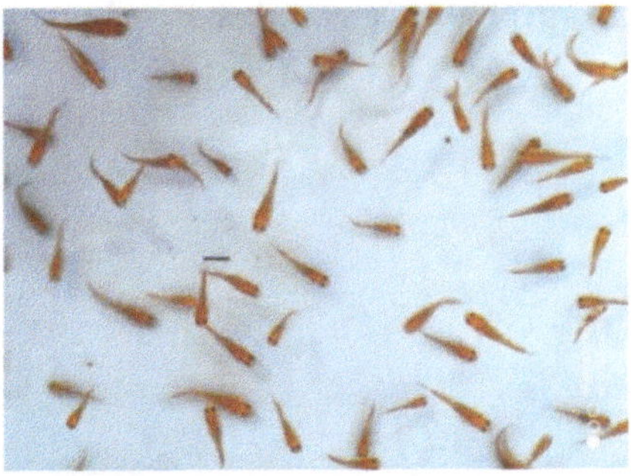

Fig. 38: Fry attained deep red colouration

Chapter 7: Fingerling Production

Fig. 40: A battery of hapa

Fig. 41: Cement tanks

Fig. 42: FRP tanks of 500 L

Fig. 44: Advanced fingerling ready for stocking in grow-out ponds

Fig. 45: Fry to fingerling rearing in biofloc media

Chapter 8: Grow-Out Production

Fig. 46: Earthen pond for grow-out production

Fig. 47: Cement tanks used for grow- out pro duction

Fig. 48: 5-month old striped murrel grown in cement tank

Fig. 50: A view of biofloc grow-out tanks

Fig. 52: Floating Pellet diet for grow-out

Fig. 54: Soaked artemia cyst

Fig. 55: Preparation of hypochlorite solution

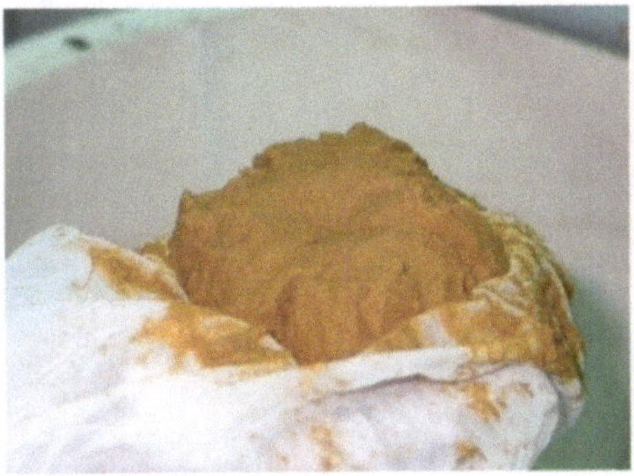

Fig. 56: Decapsulated artemia cysts

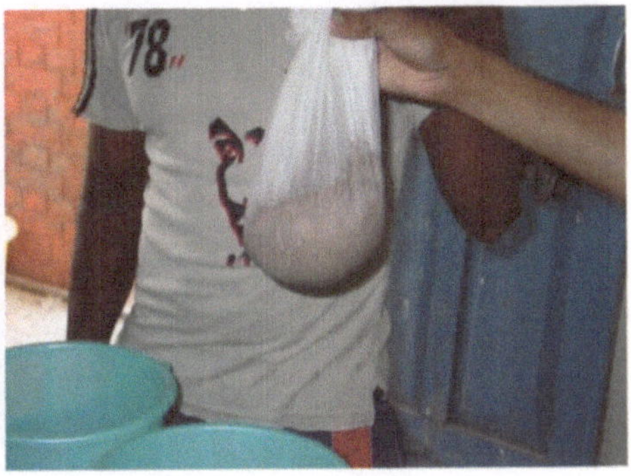

Fig. 57: Washing of decapsulated cysts

Fig. 59: Freshly hatched BSN

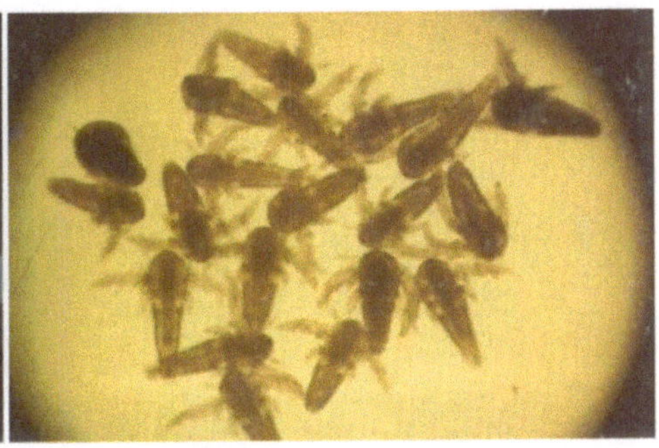

Fig. 60: BSN under microscope